AIMING FOR AN A IN A-LEVEL CHEMISTRY

Sarah Longshaw

HODDER
EDUCATION
AN HACHETTE UK COMPANY

Although every effort has been made to ensure that website addresses are correct at time of going to press, Hodder Education cannot be held responsible for the content of any website mentioned in this book. It is sometimes possible to find a relocated web page by typing in the address of the home page for a website in the URL window of your browser.

Hachette UK's policy is to use papers that are natural, renewable and recyclable products and made from wood grown in sustainable forests. The logging and manufacturing processes are expected to conform to the environmental regulations of the country of origin.

Orders: please contact Bookpoint Ltd, 130 Park Drive, Milton Park, Abingdon, Oxon OX14 4SE. Telephone: (44) 01235 827827. Fax: (44) 01235 400401. Email: education@bookpoint.co.uk. Lines are open from 9 a.m. to 5 p.m., Monday to Saturday, with a 24-hour message answering service. You can also order through our website: www.hoddereducation.co.uk

ISBN: 978 1 5104 2953 6

© Sarah Longshaw 2018

First published in 2018 by
Hodder Education
An Hachette UK Company
Carmelite House
50 Victoria Embankment
London EC4Y 0DZ

www.hoddereducation.co.uk

Impression number   10 9 8 7 6 5 4

Year   2021

Typeset by Integra Software Services Pvt. Ltd., Pondicherry, India.

Printed in India

A catalogue record for this title is available from the British Library.

Contents

Getting the most from this book

Aiming for an A is designed to help you master the skills you need to achieve the highest grades.

The following features will help you get the most from this book:

Learning objectives

> A summary of the skills that will be covered in the chapter.

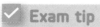 **Exam tip**

Practical advice about how to apply your skills to the exam.

Activity

An opportunity to test your skills with practical activities.

! Common pitfall

Problem areas where candidates often miss out on marks.

The difference between...

Key concepts differentiated and explained.

Annotated example

Exemplar answers with commentary showing how to achieve top grades.

Worked example

Step-by-step examples to help you master the relevant skills needed for top grades.

Take it further

Suggestions for further reading or activities that will stretch your thinking.

You should know

> A summary of key points to take away from the chapter.

About this book

Chemistry is the study of matter, including how substances combine to make new ones, how they can be separated and how they interact with energy. It can broadly be divided into organic, inorganic and physical chemistry, and it involves a practical component. As you study towards your chemistry A-level, there will be topics that you find more interesting than others and skills that you find it easier to master, but to reach an A grade, you need to be able to master them all.

To achieve an A you have to not only work hard, but you need to work effectively. This means that you need to direct your effort to the areas where it will have most impact. In addition, you need to understand where and what your weaknesses are, and address those weaknesses by applying lots of practice.

Using this book

This book is intended to guide you through the skills that you will require to reach the highest grades and help you critically evaluate and improve your own performance. Each chapter focuses on a different skill, and uses some subject content to exemplify how that skill is developed. Chemistry is synoptic, meaning there are links between the different elements of the subject; the more you practise thinking about how ideas link together, the easier you will find it to apply your knowledge.

→ Chapter 1 (quantitative skills) looks at the use of mathematical skills in the interpretation of data to support your understanding of chemistry. It will enable you to develop your ability to use maths as a tool for problem solving.

→ Chapter 2 (reading skills) offers suggestions on the best ways to extract information from the variety of sources you will encounter in your course and will also help to guide and support you in areas of your own particular interest.

→ Chapter 3 (writing skills) addresses the ability to express yourself clearly and accurately in a variety of ways, so that the information you present is accurate and logical. It will help you to establish writing habits that will make it easier for you to achieve the higher marks.

→ Chapter 4 (practical skills) will not only focus on the skills you learn and use when carrying out practical activities in chemistry, but it will also help you to apply these skills to develop a greater understanding of the design, purpose and limitations of practical work. In addition, it will enable you to consider how you use the results generated by practical work to further develop your understanding of chemical concepts.

→ Chapter 5 (revision skills) provides suggestions on how to prepare for the exam, by building good practice into your everyday experience. It recognises that cramming is unlikely to result in you achieving a top grade, and so it helps you to structure your learning from the outset of the course.

→ The Exam board focus section provides some final exam-board-specific revision advice and looks at the unique characteristics of the AQA, CCEA, Edexcel, Eduqas, OCR and WJEC courses and assessments, with an eye on achieving the highest grades.

Use this book alongside your notes and other resources; complete the activities, annotate the text and revisit particular sections as and when you need to.

The exam, assessment objectives and tips for success

A-grade students are characterised by their thorough understanding of the subject:

→ They are able to see how the different elements of chemistry are related and where there are patterns forming (synoptic thinking).

→ They are able to explain aspects of chemistry by relating back to fundamental principles.

→ They are competent at using mathematical procedures in a chemistry context and they understand the nature of scientific experimentation.

→ They understand the need for models to help develop understanding and are able to evaluate the strengths and limitations of these models.

→ They are inquiring, and constantly challenge their own knowledge and understanding by asking why things happen, or what the result would be if a change were applied, or how altering a factor would change a particular outcome.

→ They are problem solving; able to apply knowledge and principles to something they may not necessarily have studied before.

The inquisitive nature of an A-grade student may take them beyond the exam specification, but they will understand when and how to use the extended knowledge they gain.

An A-grade student is one of a small percentage of students who are awarded a higher number of marks for the qualification taken. To identify the required marks, the examining bodies make a prediction, based on prior attainment, of the expected percentage of students who will achieve the top grades. (The percentages for those achieving A grades and A* grades will be different, but the principle is the same.)

Ofqual provides guidance on the awarding process — through blogs and a range of documents — as do many of the examining bodies. More information on how the different bodies award the highest grades is found in the Exam board focus section.

Activity

Download the specification you will be using and print it off. This should become a working document where you track the coverage of topics, as well as understanding how they link together. It will also give an indication of the level of detail that will be required by your particular awarding body.

Activity

Take a topic — for example, acids — and write this in the centre of the page. Then add in all the facts you know about that topic. Add in connections to other topics, where appropriate. The ability to link ideas will help you to appreciate the synopticity of the subject as well as understanding how the questions could be structured. You could also add practical and mathematical connections. Acids, for example could be linked to organic chemistry, titrations and pH calculations.

The exam

The purpose of the exam is to find out what you know and understand. Contrary to what many students believe, exams have not been designed to catch you out; they are more of an opportunity for you to demonstrate your capability within a particular subject. Throughout the course you will have practised applying your knowledge, in the same way that athletes practise in preparation for a particular race. You will practise individual components or questions, just as an athlete may practise the start or when to accelerate, and you will also complete whole practice papers, in the same way that the athlete will run in a number of events leading up to the major competition in which they are taking part.

The exact nature of each paper will depend upon the particular awarding body, but it will cover the skills that you have developed throughout the course.

Shorter-answer questions will mostly focus on what you can recall and rely on the fact that you are expected to learn a certain number of facts and definitions. They may also test your knowledge of key mathematical and practical skills. Longer-answer questions, will require you to organise your thinking and to structure your responses. They may also draw on mathematical and practical components of the course.

Take it further

When you have learned a topic you need to apply your knowledge and assess your understanding. Once you have completed a question (or paper) and used the mark scheme to correct it, read the examiners' report. This will give you an indication of common pitfalls for each question or paper, and how to avoid them. Use what you have found out to help you improve. Analyse your mistakes and why you made them; then find other, similar questions and track your improvement.

Assessment objectives and core skills

As an A-grade student you will be aware that each question is structured to address particular assessment objectives. There are three of these for A-level chemistry, as shown in the table below:

Assessment objectives for A-level chemistry

Assessment objective	What it entails	Percentage of A-level
AO1	Demonstrate knowledge and understanding of scientific ideas, processes, techniques and procedures	30–35
AO2	Apply knowledge and understanding of scientific ideas, processes, techniques and procedures: • in a theoretical context • in a practical context • when handling qualitative data • when handling quantitative data	40–45
AO3	Analyse, interpret and evaluate scientific information, ideas and evidence, including in relation to issues, to: • make judgements and reach conclusions • develop and refine practical design and procedures	25–30

The difference between...

The table below shows the difference between core and advanced skills.

Core skills (AO1)	Advanced skills (AO2 and AO3)
• These are the more straightforward skills. • They include recalling definitions, describing procedures and carrying out calculations.	• These are more complex skills and require a degree of problem solving. • They may involve applying your knowledge, evaluating a procedure or carrying out more complex calculations including the selection and rearrangement of equations.

Activity

Locate and download the command words that your exam board uses for chemistry. Where possible, annotate these with the assessment objective that they are most likely to be associated with. For example, the command word *define* is associated with AO1. Note that it is not always possible to assign assessment objectives because in some instances, the context will determine whether the term refers to AO2 or AO3.

You should know

> To achieve the top marks you need to be proficient in all three assessment objectives.

> To progress throughout the course, you need to learn from your mistakes.

> To develop a real understanding of chemistry, you need to constantly make connections and links between different topics.

1 Quantitative skills

Study skills

Quantitative skills are integral to A-level chemistry — in the current specifications, 20% of the examination marks are for maths skills at level 2 or higher (this is equivalent to higher-tier GCSE maths). Other mathematical skills, below level 2, will not be judged as contributing to the 20% but may still be assessed — for example, the simple substitution of data in completing calculations.

The quantitative skills assessed in your specification are divided into the following areas:

→ Arithmetic and numerical computation
→ Handling data
→ Algebra
→ Graphs
→ Geometry and trigonometry

This chapter will address each of the above in turn, initially focusing on the core quantitative skills and then moving on to consider how to develop the higher-order thinking skills of analysis, interpretation and evaluation, in relation to these.

Activity

Complete and tick the activities in Table 1.1 to help you understand the main mathematical skills you will need.

The answers for activities in this chapter can be found on page 91.

Table 1.1 Mathematical skills for A-level chemistry

Mathematical skill	What you need to be able to do	Activity	Done/✓
Use of your calculator	Convert between decimals and standard form Enter numbers in standard form Use the log functions, including $\log_{10}$ and $\log_e$ (usually denoted as ln)	Make sure you are familiar with how your calculator works	
Use of the data sheet	Recognise and utilise the information given on the sheet to help you answer exam questions	Print off the data sheet for your exam board and familiarise yourself with the constants, data and equations provided	
Units	Have an understanding of which units are used to measure different quantities and be able to convert between them For example: giving units for K_c, K_p or a rate constant and converting entropy units to use in free-energy calculations	Give the units for K_c for each of the following equilibria: (a) $H_2(g) + I_2(g) \rightleftharpoons 2HI(g)$ (b) $N_2(g) + 3H_2(g) \rightleftharpoons 2NH_3(g)$	
Decimal places	Use an appropriate number of decimal places	Calculate the pH of a solution that contains $0.136\,mol\,dm^{-3}$ H^+ ions	
Standard form	Be able to write numbers in the form $a \times 10^b$ where a is an integer between 1 and 10 and b is a whole number	Write the following in standard form: (a) 0.00045 (b) 375 000	
Ratios, fractions and percentages	Be able to calculate percentage yield or atom economy Use ratios and/or percentages to derive empirical formulae	(a) 2-hydroxybenzoic acid $C_7H_6O_3$ is used to make aspirin, $C_9H_8O_4$. If a starting mass of 2.00 g of the acid is used, what is the maximum possible yield of aspirin? (b) A yield of 66.7% is achieved. What mass of aspirin does this represent? (c) 100 g of an organic compound contains 51.6 g of oxygen, 9.1 g of hydrogen, and carbon. Calculate its empirical formula.	
Estimating	Use estimates to check calculations — for example, if you are carrying out a calculation and you get a value of $0.1\,cm^3$ for a titre you would know you had gone wrong somewhere Estimate the effect of changing temperature, for example on K_c	Sulfuric acid can be neutralised using sodium hydroxide, as shown: $H_2SO_4 + 2NaOH \rightarrow 2H_2O + Na_2SO_4$ If you pipetted $25.0\,cm^3$ of $0.50\,mol\,dm^{-3}$ acid into a conical flask, what sort of range of titres would you expect if the concentration of the hydroxide was the same as that of the acid?	
Means	Calculate the weighted mean or select appropriate data values to use to calculate a mean titre	A student obtains the following values when carrying out a titration. Find the mean titre. <table><tr><td>Titration</td><td>1</td><td>2</td><td>3</td><td>4</td><td>Mean</td></tr><tr><td>Titre/cm³</td><td>22.00</td><td>22.35</td><td>22.20</td><td>22.45</td><td></td></tr></table>	
Uncertainty	Identify uncertainties in measurements	The uncertainty associated with a measuring cylinder is $\pm 1\,cm^3$. Calculate the percentage error when measuring $23\,cm^3$ of a solution with this cylinder.	

Mathematical skill	What you need to be able to do	Activity	Done/✓
Equations	Be able to change the subject of an equation as well as being able to substitute values into an equation and calculate a given quantity Be able to solve algebraic equations	The K_a of methanoic acid is $1.6 \times 10^{-4}\,mol\,dm^{-3}$. What is the pH of a $0.01\,mol\,dm^{-3}$ solution of the acid?	
Logarithms	Use logs to represent quantities that change over several orders of magnitude Find the log of a given number and, conversely, if given a log, you need to be able to find the number using the inverse or antilog	(a) What is the log of 0.25? (b) What is ln of 0.25? (c) If $\log_{10} x = 0.32$ what is x?	
Graphs	Plot graphs when given data values, ensuring that the axes and scale are correctly chosen and labelled Recognise common graphs Determine the slope and intercept for a given line graph or draw a tangent to a curve and calculate the gradient (including the units)	What order of reaction does the rate graph in Figure 1.1 represent? **Figure 1.1**	
Geometry and trigonometry	State the angles associated with particular 3D shapes Represent different 3D shapes on paper	(a) Draw a diagram to show the shape of a molecule of methane, CH_4. (b) Give a value to the bond angle.	

! Common pitfall

Students often lose marks because they give units to relative formula masses — remember that M_r values do not have units because they are comparative. Molar mass does have units (because it measures the mass of one mole of a substance).

✓ Exam tip

Remember that in Gibbs free energy calculations where you are using entropy (J) and enthalpy (kJ), you will first need to check that the units are common and convert units if necessary. Usually you convert J to kJ, but check what the question requires.

! Common pitfall

Students often lose marks if they give an answer to more decimal places than the piece of equipment can accurately measure. If you are combining different measurements, then an answer should be given to the number of significant figures justified by the least accurate piece of equipment — the accuracy of the final answer can be no greater than the least accurate measurement.

! Common pitfall

Students often lose marks when they use values other than the concordant titres (those within $0.10\,cm^3$) in their calculation of a mean titre.

Maths is a tool to support the quantitative nature of chemistry, where we might want to know how much of one chemical reacts with a given amount of another chemical, and how much energy is released when particular bonds within the product are made. However, it is important to recognise that it is a tool and not to lose sight of the context in which it is being applied. This is particularly important when you are faced with an unfamiliar scenario.

As an A-grade student you will be able to identify and use information given in the question and apply your problem-solving skills, setting out your working logically and in an easy to follow manner (see worked example 1.1). This makes it easier for you to check your working and also for examiners to credit error carried forward (ecf) marks if you do happen to make an error in a calculation.

Arithmetic and numerical computation

This skill is largely concerned with finding a value and expressing it in the correct form, to an appropriate number of significant figures and with the correct units. As an A-grade student you will be able to structure such calculations where necessary.

Worked example 1.1

Setting out your work

A student needed to prepare a buffer solution of pH 4.00 from $100\,cm^3$ of $0.500\,mol\,dm^{-3}$ ethanoic acid and a mass of sodium ethanoate. Calculate the mass of sodium ethanoate needed. (The K_a of ethanoic acid is $1.70 \times 10^{-5}\,mol\,dm^{-3}$.)

This is an example of a buffer calculation, which can be broken up into steps. First identify the information given in the question and the equation linking the species.

The information given is K_a of the acid, the concentration of the acid and the pH. The volume of acid used is also given which is needed later in the calculation.

Including the units in the calculations helps you to identify what the units should be at the end, because you can see where the units cancel. This also helps to identify possible errors in the calculation; if the units at the end are not the correct ones for the quantity you are finding, go back and check your working.

➡

Step 1: Write the equation to link K_a and the concentrations of the species present:

$$K_a = \frac{[A^-][H^+]}{[HA]}$$

where HA is the weak acid (in this case ethanoic) and A^- is the salt of the weak acid (in this case sodium ethanoate)

Step 2: Rearrange the equation to make $[A^-]$ the subject:

$$[A^-] = \frac{K_a[HA]}{[H^+]}$$

Step 3: The pH value is given, so use $pH = -\log_{10}[H^+]$ to calculate $[H^+]$:

$$pH = -\log_{10}[H^+]$$

$$4.00 = -\log_{10}[H^+]$$

$$[H^+] = 10^{-pH}$$

$$[H^+] = 1 \times 10^{-4}\,mol\,dm^{-3}$$

Step 4: Substitute the values for $[H^+]$, $[HA]$ and K_a into the equation:

$$[A^-] = \frac{1.700 \times 10^{-5}\,mol\,dm^{-3} \times 0.500\,mol\,dm^{-3}}{1 \times 10^{-4}\,mol\,dm^{-3}}$$

$$= 0.0850\,mol\,dm^{-3}$$

Step 5: The question asks for the mass of sodium ethanoate; so far a concentration has been calculated. The volume of acid ($100\,cm^3$) is the volume that the sodium ethanoate has been dissolved in (remember that a buffer contains an acid and a salt of that acid). Before calculating the number of moles of sodium ethanoate, first convert the volume from $100\,cm^3$ to dm^3, by dividing the volume by 1000.

$$\text{number of moles } (n) = \frac{\text{concentration } (C) \times \text{volume } (V)}{1000}$$

$$n = \frac{C \times V}{1000}$$

$$= \frac{0.085 \times 100}{1000}$$

$$n = 0.00850\,mol$$

Step 6: Now calculate the mass of sodium ethanoate required using the equation:

$$\text{number of moles } (n) = \frac{\text{mass}}{M_r}$$

First rearrange to make mass the subject of the equation:

$$\text{mass} = M_r \times n$$

The formula of sodium ethanoate is CH_3COONa so, using your data sheet, the M_r can be calculated:

$$(2 \times 12.0) + (3 \times 1.0) + (2 \times 16.0) + 23.0 = 82.0$$

Substituting:

$$mass = M_r \times n$$

$$= 82.0 \times 0.00850 = 0.697 \, g$$

Handling data

Significant figures

Many students struggle with the use of significant figures. As an A-grade student you will know to round your answer to a calculation to the same number of significant figures as the data value with the fewest significant figures that is used in the calculation (unless otherwise specified). This means that when measurements are multiplied or divided, the answer can contain no more significant figures than the least accurate measurement. It is also important to remember that rounding off an answer to the appropriate number of significant figures should only be done at the end of a calculation.

To find the number of significant figures, start counting from the first non-zero digit and continue up to the last non-zero digit unless there is a decimal point, in which case keep counting to the end. For example: 2780 has three significant figures but 278.0 has four.

Zeros within a number are significant, but those at the start of a number are not.

Decimal places

If you are asked to give your answer to three decimal places, look at the digit in the fourth decimal place. If it is 5 or more round the number in the third decimal place up; if it is less than 5 leave the number in the third decimal place as it is. For example:

Give 2.792103 to three decimal places.

The number in the fourth decimal place is 1, which is less than 5, so the answer is 2.792.

In the case of 2.792703, the number in the fourth decimal place is 7, which is greater than 5, so the answer is 2.793 to three decimal places

Algebra

Algebra is all about the use of equations and formulae, which may feature symbols as well as numbers and letters.

> **Activity**
>
> Give the following values to three significant figures:
> (a) 376 245 Pa
> (b) 42.000 g
> (c) 0.00732 mol
> (d) 5760 J

> **Activity**
>
> Many of the exam boards have guidance on the maths skills required. Go to your exam board's website and search the resources to find the appropriate maths skills documents. Download these and refer to them as necessary.

Activity

Complete Table 1.2, which summarises the most commonly used symbols and their meanings.

Table 1.2

Symbol	Meaning
	A reaction that is reversible because it can proceed in both directions
$\geq$	
$\approx$	
	Less than
	Approximately
$\propto$	

You need to be able to rearrange equations, including those that feature powers, logs and fractions. (Remember that whatever you do to one side, you must do to the other.)

You will come across the use of various constants in A-level chemistry. For example, in a reaction where chemicals A and B react to form a product, then the rate of the reaction may or may not vary with the concentration of A (and/or B). This can only be determined experimentally. However:

→ If the concentration of A is doubled, and the rate of the reaction also doubles, then the rate of the reaction is proportional to the concentration of A:

rate $\propto$ [A]

→ If the rate of the reaction quadruples when the concentration of B is doubled, then the rate is proportional to B^2:

rate $\propto [B]^2$

→ Since the concentrations of both reactants influence the rate:

rate $\propto [A][B]^2$

To simplify things a proportionality constant, k — the rate constant — is used and the equation is written as:

rate $= k[A][B]^2$

Worked example 1.2

Rearranging an equation

Ammonia is formed from the reaction between hydrogen and nitrogen gases, as follows:

$$N_2(g) + 3H_2(g) \rightleftharpoons 2NH_3(g) \qquad \Delta H = -92.4 \, \text{kJ mol}^{-1}$$

The gases were heated to 650 K, at a pressure of 10 atmospheres and allowed to come to equilibrium, when a sample of the mixture was analysed. The mole fraction of hydrogen was 0.655 and that of nitrogen was 0.209. Find the mole fraction of ammonia. $K_p = 3.2 \times 10^{-3} \, \text{atm}^{-2}$

Activity

Rearrange the following equations to find the value specified.

(a) $pV = nRT$ (find V)

(b) $K_c = \dfrac{[a][b]}{[c]^2}$

(find c)

(c) $n = \dfrac{m}{M_r}$

(find M_r)

(d) $\Delta G = \Delta H - T\Delta S$ (find T)

Activity

Make a list of the other proportionality constants that you have come across (you might want to return to this as you progress through your course). Write an expression including each one.

From the value of K_p, suggest where the position of equilibrium lies and also the effect that increasing the temperature and the pressure will have on K_p.

Step 1: Write the expression for K_p:

$$K_p = \frac{p(NH_3)^2}{p(H_2)^3 p(N_2)}$$

Step 2: Rearrange the equation to make the unknown the subject — doing this in stages helps you (and the examiner) see what you have done.

$$p(NH_3)^2 = K_p(H_2)^3 p(N_2)$$

$$p(NH_3) = \sqrt{K_p(H_2)^3 p(N_2)}$$

Step 3: You know the value of K_p but now need to work out the partial pressures using:

partial pressure = mole fraction × total pressure

The total pressure is 10 atmospheres:

partial pressure H_2 = 0.655 × 10 atm = 6.55 atm

partial pressure N_2 = 0.209 × 10 atm = 2.09 atm

Step 4: Substitute the values into the equation:

$$p(NH_3) = \sqrt{3.20 \times 10^{-3} \times 6.55^3 \times 2.09}$$

$$p(NH_3) = 1.37$$

Step 5: Partial pressure = mole fraction × total pressure, so to calculate the mole fraction, divide the partial pressure by the total pressure:

$$\text{mole fraction of ammonia} = \frac{1.37}{10}$$

$$\text{mole fraction} = 0.137$$

Check whether this is a sensible value — the total of the mole fractions should be 1:

$$0.137 + 0.655 + 0.290 = 1.01$$

Step 6: To think about the relationship between the value of K_p and the position of the equilibrium, look again at the equation:

$$K_p = \frac{p(NH_3)^2}{p(H_2)^3 p(N_2)}$$

K_p is a small number, so the denominator must be bigger than the numerator and there must be more reactant than product. Therefore, the position of the equilibrium lies to the left-hand side.

➡

! Common pitfall

Students often miss out on marks because they use square brackets instead of the rounded ones. Since square brackets are used to represent concentration, this makes the examiner think that you don't understand the difference between K_c and K_p.

Step 7: What is the effect of increasing the temperature?

The question shows that ΔH has a negative value, which means that the reaction is exothermic. Increasing the temperature will favour the reverse reaction, so the value of the denominator will increase and K_p will get smaller.

Step 8: What is the effect of increasing pressure on the value of K_p?

K_p is constant at a given temperature. In this reaction:

$$K_p = \frac{p(NH_3)^2}{p(H_2)^3 p(N_2)}$$

Remember that partial pressure = mole fraction, x × total pressure P_t. Rewrite the expression as:

$$K_p = \frac{(xNH_3 P_t)^2}{(xH_2 P_t)^3 (xN_2 P_t)}$$

If we cancel terms we are left with:

$$K_p = [1/P_t^2] \times \frac{xNH_3^2}{xH_2^3 N_2}$$

K_p has to stay constant because the temperature has not changed, so in order to compensate for the change in total pressure (in this case $1/P_t^2$) the mole fractions of NH_3, H_2 and N_2 also have to change. This is reflected in Le Chatelier's principle, which states that if a system at equilibrium is exposed to change, then the system will move to restore equilibrium. In this case, if the pressure was increased, the system would move to the side with fewer gaseous moles (the right-hand side). The amount of ammonia present would increase, whilst the amount of hydrogen and nitrogen would decrease; the numerator would thus be bigger, to compensate for the increase in P_t^2 in the denominator.

Graphs

Graphs show the relationship between variables and often help in the visualisation of information. You need to be able to plot a graph and to identify which variable is the independent variable, so that you know what information goes on each axis.

You need to be able to draw a line of best fit and to determine the slope and intercept of a linear graph and to be able to draw a tangent to a curve. You should also be able to interpret a graph (remember that spectra are counted as graphs too).

 Exam tip

Remember to choose a suitable scale to fill the graph paper — or at least two-thirds of it. Label the axes of your graph, including units.

Exam tip

The values of Δy (the rise) and Δx (the run), as shown in Figure 1.2, should be given to an accuracy of at least one small square (i.e. the value read from the graph must be accurate to half a small square — because there are two of them, one at each end of the line drawn). The length of the hypotenuse of the triangle drawn to calculate the gradient should also be at least half the length of the line of best fit through the points. ➡

! Common pitfall

Students often lose marks when finding a gradient by using points that are not on the line of best fit, but which are just taken from the table of values (although these points can be used if they do lie on the line of best fit).

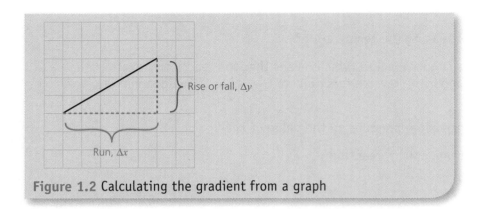

Figure 1.2 Calculating the gradient from a graph

Activity

For each of the graphs in Figure 1.3 describe the relationship between x and y and state the order of reaction given by each graph.

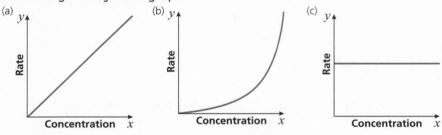

Figure 1.3

Worked example 1.3
The effect of changing variables

The Maxwell–Boltzmann distribution curve (Figure 1.4) gives an idea of the number of particles in a gas that have sufficient energy to react.

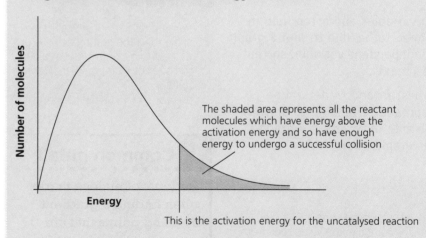

The shaded area represents all the reactant molecules which have energy above the activation energy and so have enough energy to undergo a successful collision

This is the activation energy for the uncatalysed reaction

Figure 1.4

(a) Add a dotted line to show the impact of an increase in temperature.

(b) For a given reaction the activation energy is 75.0 kJ mol⁻¹. Calculate the number of molecules in 1 mol of gas that will have an energy value greater than this when the temperature is 298 K.

(c) **Compare the impact of adding a catalyst that decreases the activation energy by 40.0 kJ mol⁻¹.**

(d) **The Arrhenius equation:**

$$k = \ln A^{\frac{-E_a}{RT}}$$

can also be written as:

$$\ln k = \ln A^{\frac{-E_a}{RT}}$$

Explain how this could be used graphically to find E_a.

Step 1: (a) Increasing the temperature shifts the curve to the right, but the peak is flatter and broader. The activation energy remains the same, but you can see that the shaded area under the curve, to the right of E_a, has now increased (Figure 1.5).

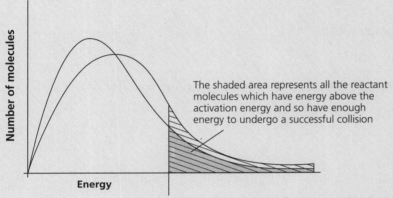

The shaded area represents all the reactant molecules which have energy above the activation energy and so have enough energy to undergo a successful collision

This is the activation energy for the uncatalysed reaction

Figure 1.5

Step 2: (b) The area under the curve represents the total number of particles. The area under the curve to the right of E_a represents the total number of particles with an energy value greater than E_a (so the total number of particles that will react).

The fraction of particles with energy greater than E_a is given by the expression:

$$e^{\frac{-E_a}{RT}}$$

From the question:

$$E_a = 75.0\, \text{kJ mol}^{-1}$$

$$T = 298\, \text{K}$$

R, the gas constant (which is given on the data sheet) $= 8.314\, \text{J mol}^{-1}\,\text{K}^{-1}$

Step 3: To get consistent units, convert the activation energy to J by multiplying by 1000:

$$E_a = 75\,000\, \text{J mol}^{-1}$$

Step 4: Substitute these values into the equation:

$$\text{fraction} = e{-}(75\,000\,\text{J mol}^{-1}/(8.314\,\text{J mol}^{-1}\text{K}^{-1} \times 298\,\text{K})$$

$$= 7.13 \times 10^{-14}$$

Notice that the units cancel.

Step 5: The total number of particles in 1 mole (Avogadro's number) = $6.023 \times 10^{23} \, mol^{-1}$, so the number of particles with an energy greater than E_a will be the fraction multiplied by the total number:

$$= 7.13 \times 10^{-14} \times 6.023 \times 10^{23}$$

$$= 4.29 \times 10^{10} \, mol^{-1}$$

Step 6: (c) Reducing E_a by 40.0 kJ mol^{-1} gives an activation energy of:

$$75.0 - 40.0 = 35.0 \, kJ \, mol^{-1}$$
$$= 35\,000 \, J \, mol^{-1}$$

At a temperature of 298 K, substituting this into the equation gives the fraction of particles as:

$$e^{\frac{-35\,000}{8.314 \times 298}}$$
$$= 7.33 \times 10^{-7}$$

And the number of particles with this energy will be:

$$7.33 \times 10^{-7} \times 6.023 \times 10^{23} = 4.41 \times 10^{17}$$

Adding a catalyst increases the number of particles with an energy greater than E_a by a factor of 10^6.

Step 7: (d) If the equation is written as:

$$\ln k = \ln A \frac{-E_a}{RT}$$

This takes the form of $y = mx + c$, which is the equation for a straight-line graph.

Plotting a graph of $\ln k$ (on the y-axis) against $\ln(1/T)$ (on the x-axis) gives a straight line with a gradient $-E_a/R$.

The value of $-E_a$ can be found by multiplying the gradient by R (the gas constant).

Geometry and trigonometry

One of the key challenges in this topic is the representation of 3D shapes on a 2D medium, such as paper, so it is often helpful to build the shapes using a molecular modelling kit to help you visualise a structure before attempting to draw it. You need to know the conventions for drawing 3D shapes, such as the use of wedges (to represent bonds coming out of the page towards you), dashed lines or reversed wedges (to represent bonds going into the page away from you) and a straight, continuous line for a bond in the plane of the paper. You also need to be able to draw the mirror image when given a particular molecule or complex.

You should be familiar with the angles associated with regular shapes. You will need to know these in order to predict the shapes of molecules by applying the electron pair repulsion theory. You must know what the original shape is before you can suggest if that will change, depending on whether there are lone pairs of electrons present or just bond pairs.

Activity

Draw the two enantiomers (optical isomers) of 2-hydroxypropanoic acid (lactic acid).

Activity

Complete Table 1.3, which summarises the different shapes and bond angles.

Table 1.3

No. of bond pairs	No. of lone pairs	Shape	Bond angle/°	Example
2	0			
3	0			
3	1			
4	0			
2	2			
5	0			
6	0			
4	2			

Worked example 1.4

Geometry

Complex ions often take part in ligand exchange reactions, many of which are accompanied by a colour change.

(a) With reference to the cobalt(II) ion, write an equation for a ligand exchange reaction in which the coordination number of the complex changes from 6 to 4.

(b) Describe the colour change(s) that you would observe during this ligand exchange and give a reason for the change in the coordination number of the complex.

(c) Draw the shape of the new complex formed.

(d) Platinum also forms complexes with a coordination number of 4, but these take on a different shape. Draw the structure of the complex cisplatin and state a use for this complex.

(e) When a ligand is added to a solution of a complex that a transition metal ion has formed with water, the ligand added exchanges with water molecules in the complex and an equilibrium is established. Write an expression for the stability constant, K_{stab} for the ligand exchange reaction in part (a).

(f) With reference to Table 1.6, explain why there is a difference in the stability constants of the complexes shown:

Table 1.4

Ligand	Complex ion	K_{stab}
Cl^-	$[CuCl_4]^{2-}$	4.0×10^5
NH_3	$[Cu(NH_3)_4(H_2O)_2]^{2+}$	1.3×10^{13}
$EDTA^{4-}$	$[Cu(EDTA)]^{2-}$	6.3×10^{18}

✓ Exam tip

The question says *with reference to the table*, so you should use information, such as the figures, from the table.

This is a question about transition metal complexes: a transition metal complex is one in which the central metal ion (in this case Co^{2+}) is bonded to a number of molecules or anions by coordinate bonds.

Step 1: (a) Write an equation for a ligand exchange in which the coordination number changes from 6 to 4.

You can choose to use either water or ammonia as the original ligand, as both are small molecules and form complexes with a coordination number of 6.

The complex with the coordination number of 4 will involve chloride ions because these are bigger. Hence:

$$[Co(H_2O)_6]^{2+}(aq) + 4Cl^-(aq) \rightleftharpoons [CoCl_4]^{2-}(aq) + 6H_2O(l)$$

Step 2: (b) Describe the colour changes you will see.

The colour changes associated with the reaction are from pink to blue so, during the transition, the colours may mix and appear a mauve or lilac colour. However, you should state the start and finish colours clearly, as well.

Step 3: (b) Give a reason for the change in shape of the complex.

The reason that the complex is a different shape is because the chloride ion is a bigger ligand than the water molecule, so it takes up more space and this means that fewer can be accommodated around the central transition metal ion.

Step 4: (c) Draw the new complex.

The new complex formed is a tetrahedral shape (Figure 1.6).

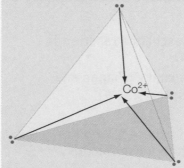

Figure 1.6

Step 5: (d) Draw cisplatin and state what it is used for.

The other shape associated with complexes with a coordination number of 4 is the square planar shape. This is the shape that platinum complexes often form due to the size of the d-orbitals (but this is beyond the scope of the A-level specification).

The formula for cisplatin is $PtCl_2(NH_3)_2$ and the prefix *cis* tells you that this is a geometric isomer and that the two ammonia ligands (and by implication, also the two chloride ligands) are next to each other. This is shown in Figure 1.8.

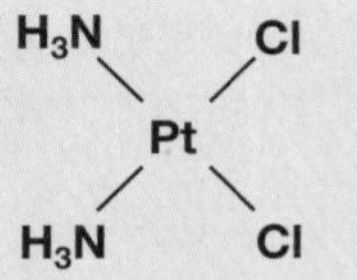

cis-platin is neutral and so diffuses through the cell membrane

Figure 1.7 Cisplatin

Cisplatin is an anticancer drug, which binds to DNA, preventing the replication of the cancerous cells.

Step 6: (e) Write an expression for the stability constant for the ligand exchange reaction in (a).

Referring back to the equation for the ligand exchange reaction:

$$[Co(H_2O)_6]^{2+}(aq) + 4Cl^-(aq) \rightleftharpoons [CoCl_4]^{2-}(aq) + 6H_2O(l)$$

$$K_{stab} = [CoCl_4^{2-}]/([Co(H_2O)_6^{2+}][4Cl^-])$$

Step 7: (f) With reference to Table 1.6 explain the differences in the stability constants of the different complexes.

Looking at the table, there are three different ligands, but Cl^- and NH_3 are both monodentate ligands, whereas EDTA is a hexadentate or multidentate ligand. This question is about entropy — the tendency of a system towards chaos or an increase in disorder. So, here the question wants you to recognise that if you swap a monodentate ligand for a multidentate one, then more molecules of water will be released (from the complex), which increases the disorder of the system and its entropy, which increases the stability of the complex.

Your answer should state that the stability of the complexes increases from the complex involving chloride, to ammonia, to EDTA, as is evident from the fact that the complex with EDTA has a stability constant of 6.3×10^{18}, whereas that with the chloride ligand has a K_{stab} of 4.0×10^5.

> **! Common pitfall**
>
> Students often lose marks by including the water molecules. However, as the reaction is carried out in aqueous solution and water will be in excess, it is never included in the expression.

Analysis, interpretation and evaluation

Analysing, interpreting and evaluating are all higher-order skills that may require you to bring together information from a variety of sources, in order to reach a conclusion.

The difference between...

Analysis	Interpretation	Evaluation
Involves examining the data in order to reach a conclusion — for example, it could involve examining the results of adding different reagents to an unknown solution in order to identify an unknown.	Involves explaining the meaning of something — for example, it could involve stating the order of a reaction by looking at how the concentration of a reactant changes with time.	Involves looking at the impact of change on a particular variable — for example, if the pressure on a gaseous system is changed, does the system move in response to that change? It also involves considering methodology and whether an experimental procedure could be improved so that the data generated are of a better quality.

Analysis

Worked example 1.5

Analysis of information

The proton NMR spectrum of compound A is shown in Figure 1.8.

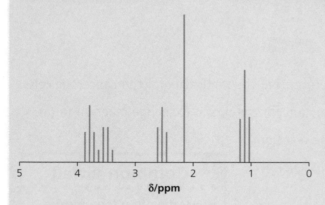

Figure 1.8

The compound has the formula $C_6H_{12}O_2$ and gives a yellow precipitate when reacted with 2,4-dinitrophenylhydrazine, which indicates that the carbonyl group, C=O, is present. The mass spectrum shows a peak at 116, a small peak at 117 and further peaks at 29 and 43.

The IR spectrum shows peaks at 1200 cm^{-1}, 1750 cm^{-1} and 2850 cm^{-1}.

Use the information above to suggest a structure for the compound, justifying how you have reached your answer.

This question asks you to *justify* your answer. To do that, you need to use your knowledge and understanding, along with all the information given in the question to provide evidence to support your answer. This is an example of where you will be using your problem-solving skills.

Step 1: Start with the atoms that are present and think about how they could be arranged.

Although there are six carbon atoms, there are too many hydrogen atoms present for the molecule to contain a benzene ring. Think about the possible functional groups that could be present. Organic molecules containing two oxygen atoms can include esters and carboxylic acids if the two oxygens are within the same functional group. Alternatively, the molecule may include two different oxygen-containing groups, such as the carbonyl or the hydroxyl group.

Step 2: As the question states that compound reacts with 2,4-dinitrophenylhydrazine, it is likely to contain the carbonyl group, in which case it should show an infrared absorption at 1640–1750 cm^{-1}, corresponding to the C=O bond, which it does. A good initial deduction is that the compound does contain the C=O group.

Step 3: Decide whether the molecule contains either the ester linkage or the carboxyl group.

The question does not mention a broad peak at 3200–3500 cm^{-1} in the IR spectrum, and there is no peak between 10 and 12 ppm on the proton NMR trace, so the molecule does not contain the carboxyl group. It is not a carboxylic acid. Sometimes information on what is *not present* can be just as useful in determining the structure as information on what is there.

➡

Step 4: Look at the mass spectrum data.

The peak at 116 will be the molecular ion peak ((12.0×6) + (1.0×12) + (16.0×2) = 116) and the small peak at 117 will be the M+1 peak (due to the presence of the carbon-13 isotope). The peak at 29 may be due to the CH_3CH_2 group and that at 43 to $CH_3CH_2CH_2$. However, the molecule also contains a C=O group, so the peak at 43 could be due to the presence of CH_3CO.

Step 5: Look at the NMR trace.

Remember that: the number of peaks gives the number of chemically different types of proton in the molecule; the shift tells us which groups are present; the integration gives us the number of each type of proton; and the splitting pattern is due to the number of chemically different protons on an adjacent carbon. The number of peaks is given by the $n + 1$ rule, where n is the number of non-equivalent neighbouring hydrogen atoms.

The peak at 1.2 has an integration value of 3, which means that there are three protons of this type and it is a triplet, which suggests that there is a CH_2 group adjacent. This is deduced from the $n + 1$ rule — if there are three peaks, then there must be two non-equivalent protons adjacent, since 2 + 1 = 3. The triplet itself will cause a quartet splitting pattern; this is present at 3.5, which is the shift due to CHO. Hence the fragment CH_3CH_2CHO is present in the molecule.

Step 6: The shift at 2.2 is a singlet, which suggests that there are no adjacent protons. It has an integration of 3, suggesting a CH_3 group. The shift at 2.6 would suggest that a C=O group is adjacent (hence the lack of splitting). This gives the fragment CH_3CO.

Step 7: The fact that there are two methyl groups could indicate the two ends of the molecule. Going back to the NMR trace, there are two triplet splitting patterns, each with an integration of 2. One of these is at shift 2.6, and so adjacent to the C=O group. The other is at 3.8, which suggests the presence of an oxygen atom or an ester group. However, you have already deduced that there is no ester group present.

Step 8: Putting all this information together, a deduction for the structure is:

$$CH_3COCH_2CH_2OCH_2CH_3$$
$$\ \ \ 1\ \ \ \ 2\ \ \ \ 3\ \ \ \ 4\ \ \ \ \ \ 5\ \ \ \ 6$$

The molecule contains the ketone group, which complies with the positive precipitation reaction with 2,4-DNPH, as well as the presence of the IR absorption.

There are five peaks in the proton NMR trace, which coincides with the fact that there are five different proton environments, as shown.

Interpretation

A question may not explicitly ask you to describe a graph, but it may expect you to use the information displayed in a graph to derive an answer. There is every possibility that it may refer to a process or reaction that you are unfamiliar with. However, the key here is to think about what the question is about and then to apply what you know to the new context. Be clear about what each part of the information presented is telling you.

Look carefully at any graphs, making sure that you understand what they show and the units used (remember to quote these to back up your answer). Look carefully at the axes of graphs to see how variables are related. Remember that you should describe how the quantity on the *y*-axis changes with respect to that on the *x*-axis.

Worked example 1.6

Using the information from a graph

Sulfuric acid is manufactured by the Contact process, the first stage of which involves the reaction of sulfur dioxide gas with oxygen to form sulfur trioxide:

$$2SO_2(g) + O_2(g) \rightleftharpoons 2SO_3(g) \quad \Delta H = -196\,kJ\,mol^{-1}$$

Figures 1.10 and 1.11 show the percentage conversion of the reactants to products under different conditions at equilibrium.

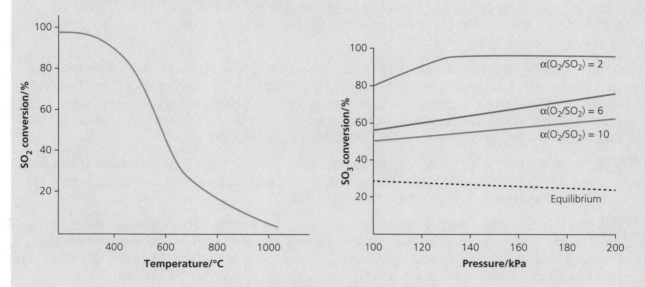

Reproduced with permission from Elsevier

Figure 1.9   Figure 1.10

Deduce the optimum conditions for the formation of sulfur trioxide from sulfur dioxide, giving reasons for your answer.

Look carefully at the command words in this and the previous example. They are similar in that they are associated with the higher-order skills of reaching conclusions from information, but they have subtly different meanings.

Step 1: *Deduce* means draw conclusions from the information given, so start by considering all the information, including that from the graphs, as well as what you know about this or similar reactions (such as the Haber process, which is also an exothermic reaction involving gaseous reactants and products).

The reaction is exothermic, so increasing the temperature will decrease the yield. However, for industrial processes a sufficiently high rate of reaction is also needed. Looking at the graph, the yield drops at about 400°C, so a temperature a little lower than this would give a high enough rate, without reducing the yield too much.

Step 2: The reaction involves gaseous reactants and the graph shows that increasing the pressure increases the percentage conversion, but again this will be limited industrially by safety considerations. Also, once the pressure reaches 130 kPa, the yield plateaus.

Step 3: Sensible conditions to suggest would thus be a temperature of 375°C and a pressure of 130 kPa.

The difference between...

Justify	Deduce
Support a case with evidence	Draw conclusions from the information given
Say why you are making deductions from each piece of information	

Evaluation

Questions involving evaluation might require you to think about how a particular factor may affect a quantitative result. This may involve thinking about how a particular procedure could be improved. For example:

→ by changing a method (e.g. so that it reduces heat loss in the case of calorimetry)

→ by using a different piece of equipment (e.g. a digital thermometer that records temperatures to the nearest 0.1°C as opposed to 0.5°C)

→ by using a larger quantity of a reagent, which will reduce the percentage error in its measurement

For a given piece of equipment, the accuracy or percentage error is fixed. For a measurement made with this equipment:

$$\text{percentage error} = \frac{\text{error uncertainty of the equipment}}{\text{actual measurement made with the equipment}} \times 100$$

The larger the denominator then the smaller the error will be. You may also be asked to think about the impact not just of using different pieces of equipment (or even a different chemical), but also what might happen in a number of different scenarios.

Application to the exam

Sample question

The following question involves a calculation, for which you need to determine the stages using the information given in the question to guide you. It then extends to test your evaluative skills, by asking you to consider the impact of two different factors. The guidance given suggests how you might approach it, and there is a sample answer with suggested mark scheme for you to compare with your response.

! Common pitfall

Students miss out on marks if they fail to use the information given to answer the question. In the previous example, failing to comment on the fact that the percentage conversion falls rapidly from 200°C to 600°C, and then more slowly, would lose marks. Similarly, the increase in pressure, from 100 kPa to 130 kPa, increases the yield from 80% to 95%, where it then plateaus.

Worked example 1.7

Propylpropanoate, $CH_3CH_2COOCH_2CH_2CH_3$ and water react in the presence of an acid catalyst, to form the alcohol and the carboxylic acid. Over time the following equilibrium is reached:

$$CH_3CH_2COOCH_2CH_2CH_3 + H_2O \rightleftharpoons$$
$$CH_3CH_2COOH + CH_3CH_2CH_2OH \quad \Delta H = -54.2\,kJ\,mol^{-1}$$

A student carried out an experiment to determine K_c for the reaction. $20\,cm^3$ of the ester (0.201 mol) and $5\,cm^3$ of $2\,mol\,dm^{-3}$ HCl (the catalyst for the reaction) were mixed together. In the mixture there were 0.730 mol of water. The mixture was left for a week to reach equilibrium at room temperature and pressure. The total volume of the equilibrium mixture was $25\,cm^3$.

(a) After a week, the student titrated $5\,cm^3$ aliquots of the equilibrium mixture against $0.75\,mol\,dm^{-3}$ NaOH. The mean titre required to neutralise this solution was $31.8\,cm^3$. Show how the student used this information to find K_c. (6)

(b) The student's value for K_c was different from that given in a data book. A possible reason is that the NaOH used might have been $0.5\,mol\,dm^{-3}$ (instead of $0.75\,mol\,dm^{-3}$). How would this affect K_c? Give a reason for your answer. (3)

(c) The student left the equilibrium mixture close to the radiator, so the temperature could have risen to more than 25°C. How might this affect K_c? Again, give a reason for your response. (3)

How to approach the question

Read through the question carefully, annotating it with any information that you will need in the calculation. Identify where you have a concentration and a volume of a particular reagent — this will be your starting point. Write down the formulae and equations you will need.

Annotated example

(a) number of moles of NaOH = (volume × conc)/1000 = (31.8 × 0.75)/1000 = 0.02385 mol ✓

Correctly calculates the number of moles of alkali required to neutralise the acid.

This is in a $5\,cm^3$ aliquot from $25\,cm^3$ of solution.

total number of moles of acid = 5 × 0.02385 = 0.11925 mole ✓

Correctly calculates the total number of acid moles in the equilibrium mixture.

The acid is made up from the propanoic acid *plus* the HCl catalyst.

number of moles of catalyst = (volume × conc)/1000 = (5 × 2)/1000 = 0.01 mol

number of moles of propanoic acid = 0.11925 − 0.01 or 0.10925 mol ✓

Correctly calculates the number of moles of propanoic acid in the equilibrium mixture (moles of propanoic acid minus moles of HCl (catalyst).

From the stoichiometry of the equation:

$$CH_3CH_2COOCH_2CH_2CH_3 + H_2O \rightleftharpoons CH_3CH_2COOH + CH_3CH_2CH_2OH$$

1 mol of acid is generated from 1 mol of ester. So:

number of moles of ester at equilibrium = 0.201
− 0.10925 = 0.09175 mol

number of moles of water at equilibrium = 0.730
− 0.10925 = 0.62075 mol ✓

Correctly calculates
the number of mole
of ester *and* water at
equilibrium.

Given that the ratios of the number of moles are
the same, the volumes will cancel, so the number
of moles can be used in place of concentration.

$$K_c = \frac{[\text{products}]}{[\text{reactants}]} =$$
$$\frac{[CH_3CH_2CH_2OH][CH_3CH_2COOH]}{[CH_3CH_2COOCH_2CH_2CH_3][H_2O]}$$ ✓

Identifies the correct
equation for K_c for
this reaction.

$$K_c = \frac{0.10925^2}{0.09175 \times 0.62075} = 0.210$$ ✓

Calculates the correct value for Kc, with no units
because in this case they cancel:

$$K_c = \frac{[0.10925/0.025][10925/0.025]}{[0.09175/0.025][0.602075/0.025]}$$

= 0.210 (calculator
value 0.209565645)

Here the student
has shown all
their working and
achieves full marks
for correct working.

The value is rounded up at the end of the
calculation to an appropriate number of significant
figures, as indicated in the question. This is
determined by the limits of the least accurate
measurement.

Correctly describes the
impact on the value of
K_c. The values in the
numerator will thus be
lower and so K_c will be
smaller.

(b) If the student uses NaOH of a lower concentration, then the number of moles of alkali will be
2/3 that calculated. ✓

Recognises that the
number of moles of
NaOH will be lower.

This means that there will be fewer moles of propanoic acid in the equilibrium mix ✓ (the amount
of HCl will be the same).

Links the number of
moles of alkali to a
smaller number of
moles of propanoic acid
(product).

This then means that there will also be less
propanol made and so there will be more reactant
left. The denominator of the equation will be bigger and so the value of K_c will be lower. ✓

Recognises that the
forward reaction is
exothermic.

(c) K_c can be affected by temperature. An increase in temperature will shift the equilibrium
in the direction of the endothermic reaction ✓,
which in this case is towards the reactants.

Correctly describes
the impact of this
change on K_c — the
amount of reactants
will be greater, so
the value of K_c will
again be lower.

States that the increase
in temperature will
favour the production
of reactants.

If this happens, there will be less product ✓ made
and so the numerator will be lower and the value
of K_c will be less ✓.

Exam tip

Do not assume that because the answer to one part of the question will lower the value of K_c, the answer to the other will raise it.

Take it further

The University of Birmingham and the University of Leeds have created a Maths for Chemists book, which provides further exemplification of some of the concepts met at A-level and beyond. Although aimed at undergraduates studying chemistry, there are some interesting and useful worked examples that are equally applicable to A-level. Find it for free via the Royal Society of Chemistry website (http://rsc.org).

You should know

> **Which mathematical skills will be assessed at A-level and how to answer questions incorporating these skills.**

> **How to estimate an answer, including the use of units to help confirm correct procedure.**

> **How to describe and explain quantitative data using your scientific knowledge**

> **How to interpret data presented in graphs and charts, including spectra, and use them to draw conclusions.**

2 Reading skills

Learning objectives

> To understand how to get the most out of your textbook.

> To develop your reading skills so that you can read more effectively.

> To understand how to extend your knowledge by reading beyond the specification.

> To develop critical thinking skills.

Study skills

Developing your reading skills is important, not only for your A-level years but also beyond, when you may go on to further study and will need to read a wider variety of materials.

Reading serves a number of purposes at A-level: it provides a background to what you are studying; it enables you to follow areas that you are particularly interested in; it describes how to carry out a practical investigation; and it tells you what you need to do when being assessed.

There are different sources and formats of information that you can read. These include:

→ textbooks
→ websites
→ journals
→ scientific papers
→ popular science books

Getting the most out of your textbook

Most schools or colleges will either provide you with, or recommend, a textbook to accompany your course.

Building on prior knowledge

Many topics that are covered in the first year of the course are revisited and extended in the second year. Before you start a topic, you should remind yourself what you already know about it.

Activity

Choose a topic and write a list of the key points. Try to do as much as you can from memory before referring to your notes. If your textbook has a checklist, compare your list with that.

If your textbook also includes short recall questions, you should complete these as a starting point to the topic.

If there are things that you do not remember or questions that you got wrong, you need to go back over these topics.

It is good practice to read ahead, so that you already have an idea of the key points of a topic before it is covered in class. This will also enable you to identify any questions that you want answered — take these to your lesson and tick them off as they are covered. Any that remain after the lesson can then be addressed by further self-study or by asking your teacher for clarification. Some teachers will give you a schedule of what they are covering and when, but you should have your own copy of the specification to help guide you through a topic.

Reading for understanding

If you own your textbook, you will be able to highlight it; if it has been provided by the school, you could use post-it notes. You will probably find that it helps to summarise your reading as you progress through a topic. This may be by recording key definitions onto cue cards (see chapter 5: Revision skills), or producing a mind map or summary of the information you have read. Flow diagrams can be helpful for summarising processes and 'dual coding' — where you use illustrations and words — is another useful method for helping you to remember information.

The passage in Figure 2.1(a) describes how a heterogeneous catalyst works, using the example of iron in the manufacture of ammonia by the Haber process. Figure 2.1(b) gives an example of how dual coding can be used to help you understand and remember this process by using diagrams to replace larger quantities of text. Completing a diagram like this means that you are creating your own interpretation and therefore are more likely to recall and to be able to apply the logic to other industrial processes.

(a)

Tip

Make sure you don't confuse adsorb with absorb.

Heterogeneous catalysts

A **heterogeneous catalyst** is in a different phase from the reactants. The most common type of heterogeneous catalysis involves reactions of gases in the presence of a solid catalyst.

The mode of action of a heterogeneous catalyst is different from that of a homogeneous catalyst. A heterogeneous catalyst works by adsorbing the gases onto its solid surface. This adsorption weakens the bonds within the reactant molecules, which lowers the activation energy for the reaction. Bonds are broken and new bonds are formed. The product molecules are then desorbed from the surface of the solid catalyst.

Transition metals are often used as heterogeneous catalysts. Iron is the catalyst in the Haber process for the manufacture of ammonia (page 176). The iron is either finely divided (and therefore has a large surface area) or in a porous form containing a small amount of metal oxide promoters:

$$N_2(g) + 3H_2(g) \xrightleftharpoons{\text{Fe catalyst}} 2NH_3(g)$$

The production of poly(ethene) and other polymers from alkenes requires the use of a Ziegler–Natta catalyst. This is a mixture of titanium(IV) chloride and an organic compound of aluminium, $Al_2(CH_3)_6$. The mode of action of such catalysts is complicated and not altogether understood. Another important heterogeneous catalyst is the alloy of platinum, palladium and rhodium used as catalytic converters in vehicles.

(b)

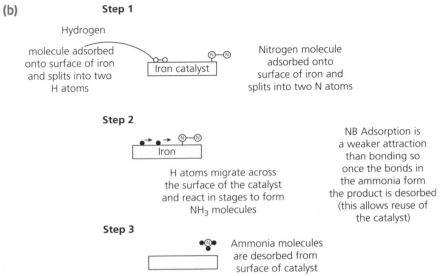

Figure 2.1 Summarising a textbook extract diagrammatically

Activity

Read the text in Figure 2.2 and summarise it using an annotated diagram of the lead–acid cell.

Lead-acid cell

The lead-acid cell is the cell used in cars and other vehicles. It is recharged as the vehicle moves. Sulfuric acid is used to provide the acid in these types of battery.

The half equations for the reaction are:

$$PbO_2(s) + 3H^+(aq) + HSO_4^-(aq) + 2e^- \rightarrow PbSO_4(s) + 2H_2O(l) \quad E^1 = +1.69\,V$$

$$PbSO_4(s) + H^+(aq) + 2e^- \rightarrow Pb(s) + HSO_4^-(aq) \quad\quad E^1 = -0.46\,V$$

At the negative electrode, oxidation occurs:

$$Pb(s) + HSO_4^-(aq) \rightarrow PbSO_4(s) + H^+(aq) + 2e^-$$

At the positive electrode, reduction occurs:

$$PbO_2(s) + 3H^+(aq) + HSO_4^-(aq) + 2e^- \rightarrow PbSO_4(s) + 2H_2O(l)$$

The overall reaction occurring when the cell is discharged is:

$$PbO_2(s) + 2H^+(aq) + 2HSO_4^-(aq) + Pb(s) \rightarrow 2PbSO_4(s) + 2H_2O(l)$$

When the cell is recharged the reaction above is reversed to regenerate the reagents:

$$2PbSO_4(s) + 2H_2O(l) \rightarrow PbO_2(s) + 2H^+(aq) + 2HSO_4^-(aq) + Pb(s)$$

At the negative electrode, Pb is oxidised from 0 in Pb to +2 in $PbSO_4$.

At the positive electrode, Pb is reduced from +4 in PbO_2 to +2 in $PbSO_4$.

The formation of lead( ) sulfate can be a problem if a lead-acid cell is discharged for long periods of time. The insoluble lead( ) sulfate build up in the cell and the cell cannot be recharged.

Figure 2.2 Extract from a textbook

Researching a topic

You may need to research a topic, perhaps for one of the required practicals, to complement your own understanding or as a learning activity. If you are using the internet for your research, defining the search is important. Start by identifying the key words you need to include. When selecting from the sources returned, bear in mind that .edu and .ac are usually associated with academic institutions and so you may find the information goes beyond the scope of the A-level specification.

Make sure you keep a note of any sources that you use, so that you are able to find them again and to reference them in your work. If you are quoting directly from any of your sources, particularly in practical write-ups, then you should indicate these sections by enclosing them in quotation marks and the source should be referenced, otherwise you could be accused of plagiarising others' work. If the source you use is a website, then you need to include the date you accessed it in your references, because website content can change over time.

Research information may involve the use of sources other than text. These include lectures and documentaries. Many of the universities have outreach programmes, which are a good way of finding out about chemistry beyond A-level.

Take it further

The University of Manchester has a dedicated YouTube channel called CAMERA (Chemistry at Manchester Explains Research Advances), which presents short insights into recent advances in chemical research in an easy-to-understand format.

Reading beyond the specification

You may find that your reading takes you beyond the exam specification and that the ideas that are presented in your reading differ from those that you encounter in A-level lessons.

The series *A Very Short Introduction*, published by Oxford University Press, includes editions focused on organic chemistry, physical chemistry, the periodic table, molecules and the elements. These books provide a further insight into different aspects of chemistry and are ideal for those who want to take their interest a little further without going into the level of detail associated with undergraduate study. One of the benefits of these bite-sized explorations is that they allow you to learn a little more about a range of different topics, rather than learning a lot about a single aspect. The content is accessible and you should also be able to recognise when it extends beyond your specification.

Exam tip

Reading beyond the specification is a great way to enhance your understanding of a topic, but remember that you need to explain concepts according to the information presented at A-level, so make sure you can do this accurately first.

There may be particular topics that interest you and that you want to find out more about, such as the periodic table and the elements. The following books are a good starting point.

➜ *Periodic Tales* — Hugh Aldersley-Williams
➜ *The Disappearing Spoon* — Sam Kean
➜ *The Secret Life of the Elements* — Ben Still
➜ *Uncle Tungsten* — Oliver Sachs

Your interests are personal to you, so commenting on why you chose to read a particular book and some of the points that appealed to you is something that you could include in your personal statement, or refer to in a university interview.

Different methods of reading

There are four main techniques of reading, and each has its own particular purpose.

→ Skimming
→ Scanning
→ Intensive
→ Extensive

Skimming and scanning are both quick methods of reading, where you are looking for information.

Skimming involves reading to get an overall impression before deciding whether to dedicate more time and attention to the text. You might skim an article to decide whether it is really telling you what the title suggests and so will be of interest or relevance to you; or you might skim a book to see if you want to buy it because it presents different information from that which you already know.

Scanning involves you looking for particular information, such as key words, to direct your attention to the exact focus for your reading. You might scan a practical method to see if it suggests quantities or whether it uses the chemicals you have selected.

Intensive reading, as the name suggests, takes more time and is used to help you master a concept. It is also likely that you will want to record information, such as definitions or equations, when reading intensively, and for this reason it is often a more active process. This type of activity is described above, in the section on reading for understanding.

Annotated example
Intensive reading

The extract in Figure 2.3 is taken from an A-level chemistry textbook. It has been annotated to suggest how you might read it intensively and what questions it might generate for further study.

You should know the definitions of key terms in the book.

Think about what functional group is present

How does nylon differ from the polymers formed from amino acids?

Are there other forms of nylon and what does the numbering mean?

The first synthetic and commercially important polyamides were various forms of nylon. These were not, however, produced from amino acids. Instead, they were formed by condensation polymerisation between diamines and dicarboxylic acids. One of the commonest forms of nylon is nylon-6,6. This is made by a condensation reaction between 1,6-diaminohexane and hexanedioic acid (Figure 18.2.25). The product is named nylon-6,6 because both monomers contain six carbon atoms.

Key terms

Polyesters are polymers with ester links between monomer units.

Polyamides are polymers with amide links between monomer units.

Tip

Early polymer chemists found it easier to synthesise polyamides using separate dicarboxylic acid and diamine molecules rather than have the carboxylic acid functional group and the amine functional group on the same molecule, as nature does in an amino acid.

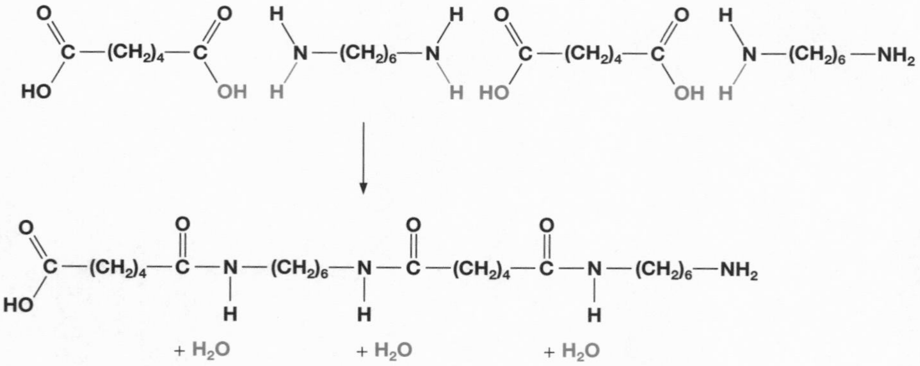

Figure 18.2.25 Condensation polymerisation to make nylon-6,6

Why is the hexanedioyl dichloride more reactive than the hexane dioic acid?

What safety precautions might you need to be aware of when carrying out this practical?

Why does the nylon form at the interface between the two solutions?

Nylon-6,6 can be produced more readily in the laboratory using hexanedioyl dichloride in place of the less reactive hexanedioic acid. A solution of hexanedioyl dichloride in cyclohexane is floated on an aqueous solution of 1,6-diaminohexane. Nylon-6,6 forms as a skin at the interface and can be pulled out as fast as it is produced forming a long thread – the 'nylon rope'.

In this reaction hydrogen chloride molecules are eliminated in the condensation reaction (Figure 18.2.26).

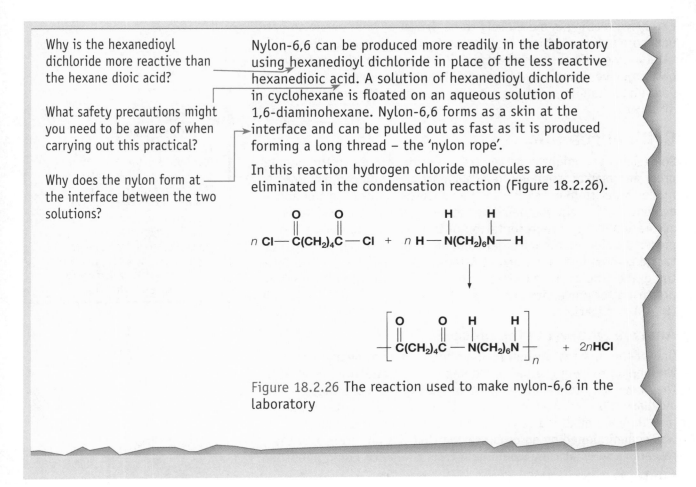

Figure 18.2.26 The reaction used to make nylon-6,6 in the laboratory

Figure 2.3 Extract from a textbook for intensive reading

Extensive reading is more likely to be driven by your desire to find out more about a subject that interests you and is also likely to take you beyond the specification. This type of reading is for your own enjoyment and so you are not likely to make notes or work at your understanding, although it may lead to further reading for discovery.

Critical reading

Sometimes you might come across scientific articles in the popular press and media. When you read such articles, you can employ higher-order thinking skills, such as analysis, interpretation and evaluation, to help you decide whether what you are reading is accurate and a fair representation of the facts presented in the piece. Developing higher-order thinking skills will help you to tackle the hardest questions at A-level, and enable you to obtain a top grade. The content of this material might also be useful when preparing for university entrance interviews or in writing university (UCAS) applications.

Analyse who wrote the piece

One of the first tasks when reading critically is to decide whether the person writing the piece is doing so independently, or whether they have an interest in the content that would make their evaluation biased. You might want to ask who the writer is and how they are qualified to write the piece, or whether the piece was commissioned by an organisation that has an interest in the research and what they would gain from the piece being published.

Evaluate where it appears

You also need to think about where the information is published. Is the site reputable? Do similar stories appear elsewhere and, if so, are the report contents similar? (It is worth also considering the credentials of these sources and their authors.) Remember that if the source is a scientific journal, then the research will have been peer-reviewed. If the article you are reading contains a link to a journal, it is worth following this to read the original piece on which the article is based.

Analyse the article

As you read the article, think about what it is saying and how it is presented. Is it easy to understand? If it has been simplified, has this been done in such a way that it is still accurate? Is the information presented as fact or opinion, or is it impossible to determine which is which? If opinions are given, who gives them and are they also credible? If the piece quotes expert opinion, then the expert should be identified.

Evaluate the conclusion

Once you have read the article, you can then decide whether you agree with the conclusions that the author makes. You may decide that the evidence would have led you to draw different conclusions.

> ## Take it further
>
> *Stuff Matters* by Mark Miodownik (2013; ISBN 978-0-241-95518-5) takes a number of everyday materials and explains their properties by dissecting them at the molecular level. Described as 'an ode to material science', this discussion of concrete, steel, glass, paper, chocolate, celluloid and so on 'makes even the most everyday (materials) seem thrilling'.

Annotated example

Critical reading

The following article, published in the *Guardian* on 27 June 2017, linked an increase in the use of paint stripper with a delay in the recovery of the hole in the ozone layer.

An internet search for an author should give some information about their background, such as their qualifications and employment history. This will help you to decide whether they have studied a subject in detail and at which institution. This will also help decide whether they are likely to be sufficiently qualified to understand the content in depth. Damian Carrington, the journalist in this case, has an MA in Earth Science from Cambridge University and a PhD in Geology from the University of Edinburgh, which suggests that he is qualified to write such an article.

An explanation of ozone and what it does.

Ozone hole recovery threatened by rise of paint stripper chemical — Damian Carrington, Environment Editor

Tuesday 27 June 2017 16.00 BST

The restoration of the globe's protective shield of ozone will be delayed by decades if fast-rising emissions of a chemical used in paint stripper are not curbed, new research has revealed.

Atmospheric levels of the chemical have doubled in the last decade and its use is not restricted by the Montreal Protocol that successfully outlawed the CFCs mainly responsible for the ozone hole. The ozone-destroying chemical is called dichloromethane and is also used as an industrial solvent, an aerosol spray propellant and a blowing agent for polyurethane foams. Little is known about where it is leaking from or why emissions have risen so rapidly.

The loss of ozone was discovered in the 1980s and is greatest over Antarctica. But Ryan Hossaini, at Lancaster University in the UK and who led the new work, said: 'It is important to remember that ozone depletion is a global phenomenon, and that while the peak depletion occurred over a decade ago, it is a persistent environmental problem and the track to recovery is expected to be a long and bumpy one.'

'Ozone shields us from harmful levels of UV radiation that would otherwise be detrimental to human, animal and plant health,' he said.

Clearly links the delay in the closing of the hole in the ozone layer to the presence of a chemical found in paint stripper.

References the work of the Montreal Protocol.

Quotes a named scientist, Ryan Hossaini of Lancaster University, as well as quoting an atmospheric physicist from Manchester University, Grant Allen, and two of the scientists from the British Antarctic Survey — Jonathan Shanklin and Anna Jones — all of which adds to its credibility.

Cites research published in the journal *Nature Communications*, which features a peer-review system, giving credibility to articles.

The new research, published in the journal *Nature Communications*, analysed the level of dichloromethane in the atmosphere and found it rose by 8% a year between 2004 and 2014. The scientists then used sophisticated computer models to find that, if this continues, the recovery of the ozone layer would be delayed by 30 years, until about 2090.

Details of how the rising levels of dichloromethane could affect the hole in the ozone layer.

The chemical was not included in the 1987 Montreal protocol because it breaks down relatively quickly in the atmosphere, usually within six months, and had not therefore been expected to build up. In contrast, CFCs persist for decades or even centuries.

Discusses the difference between CFCs and dichloromethane, and why the latter was not included in the Montreal Protocol.

But the short lifespan of dichloromethane does mean that action to cut its emissions would have rapid benefits. 'If policies were put in place to limit its production, then this gas could be flushed out of the atmosphere relatively quickly,' said Hossaini.

If the dichloromethane in the atmosphere was held at today's level, the recovery of the ozone level would only be delayed by five years, the scientists found. There was a surge in emissions in the period 2012–14 and if growth rate continues at that very high rate, the ozone recovery would be postponed indefinitely, but Hossaini said this extreme scenario is unlikely: 'Our results still show the ozone hole will recover.'

Details of how the rising levels of dichloromethane could affect the hole in the ozone layer.

Quotes an atmospheric physicist from Manchester University, Grant Allen, and British Antarctic Survey scientist Jonathan Shanklin, which adds to its credibility.

Grant Allen, an atmospheric physicist at the University of Manchester, said: 'Whatever the source of this gas, we must act now to stop its release to the atmosphere in order to prevent undoing over 30 years of exemplary science and policy work which has undoubtedly saved many lives.'

Jonathan Shanklin, one of the scientists at the British Antarctic Survey (BAS) who discovered the ozone hole in 1985, said: 'The Montreal protocol has proved very effective at reducing the emissions of substances that

Discusses the difference between CFCs and dichloromethane, and why the latter was not included in the Montreal Protocol.

Quotes another named scientist from the British Antarctic Survey, Anna Jones — which adds to its credibility.

References the work of the Montreal Protocol.

can harm the ozone layer. I am sure that the warning made in this paper will be heeded and that dichloromethane will be brought within the protocol in order to prevent further damage to the ozone layer.'

There are other short-lived gases containing the chlorine that destroy ozone, but few measurements have been taken of their levels in the atmosphere. 'Unfortunately there is no long-term record of these, only sporadic data, but these do indicate they are a potentially significant source of chlorine in the atmosphere,' said Hossaini, adding that further research on this was needed.

Anna Jones, a scientist at BAS, said: 'The new results underline the critical importance of long-term observations of ozone-depleting gases and expanding the Montreal Protocol to mitigate new threats to the ozone layer.'

Overall the Montreal Protocol is seen as very successful in cutting ozone losses, with estimates indicating that without the protocol the Antarctic ozone hole would have been 40% larger by 2013. Scientists discovered four 'rogue' CFCs in 2014 that were increasing in concentration in the atmosphere and contributing to ozone-destruction.

The article concludes by stating that the Montreal Protocol has been successful in reducing the hole in the ozone layer and that the new CFCs discovered were contributing to the reversal of this process. This is reinforced by the statement about the need for long-term observation. The conclusion is logical — it is based on statements made within the text of the article and it uses estimates rather than claiming particular values. The language mirrors that in the article — it is factual and backed by evidence.

As an A-grade student, when reading articles like this, you will recognise where the theme fits into what you have studied and you will apply your own knowledge to help you understand the implications raised in the article. You might jot down the formulae of the products mentioned, or think about how these can be made and then how they are used to make PVC, for example. You might also recall what you know about the Montreal Protocol and the catalysed breakdown of ozone by chlorine radicals.

Extract © Guardian News & Media Ltd 2018

Activity

A further search reveals a similar story in other sources, such as the *New Scientist*, June 27 2017 (www.newscientist.com/article/2138753-ozone-layer-recovery-will-be-delayed-by-chemical-leaks), as well as in other newspapers and on university websites. Produce a one-paragraph summary of the *New Scientist* article. You may wish to use the headings in Table 2.1 to help. Then check your summary against other reports and see if you want to add any additional information.

Table 2.1

Article	Identify who wrote it	Other journal references or citations	Analyse the article	Evaluate the conclusion
New Scientist, www.newscientist. com/article/2138753-ozone-layer-recovery-will-be-delayed-by-chemical-leaks				

Application to the exam

By reading a variety of sources, you will develop the ability to assimilate information more quickly and to pick out the most relevant parts. You will identify the subject and the exemplification. You will also have extended your knowledge of chemistry in a variety of different contexts, so that you are able to apply what you know about a topic in an unfamiliar context.

You will need to apply your reading skills in the exam. To secure top marks, you need to read *and* answer every part of each question.

The first skill you will use is that of skimming the question, reading it through to the end, to get an overall idea of the topic it is based on and how it develops. Having established this, then read the question intensively, highlighting the information that you will need to use to answer it.

Before you start to write your answer, scan the question for information on *how* to answer it — look for key command words and whether a particular number of significant figures or unit is required.

! Common pitfall

Many students waste valuable time in an exam giving information that is not required. Although this will not be penalised (unless it is incorrect), it may mean that marks are subsequently lost if the student runs out of time.

Worked example 2.1

The structures of salbutamol, a compound used in inhalers, and ethylamine are shown in Figure 2.4.

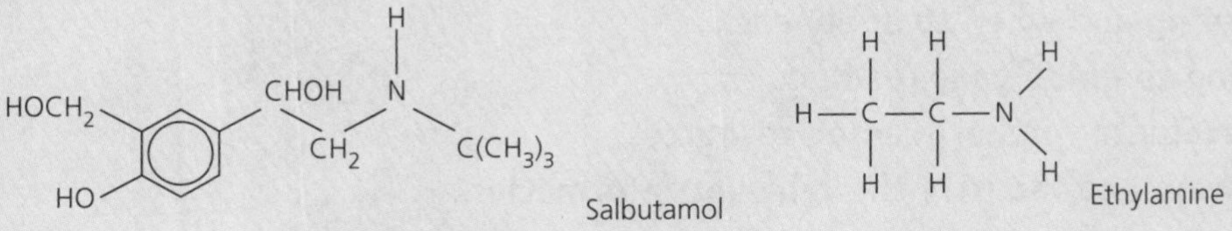

Figure 2.4

(a) State the feature that accounts for the ability of an amine to act as a base. **(1)**

(b) Describe and explain the difference in the strength of ethylamine and salbutamol as bases. **(3)**

Step 1: Skim-read the question. It is about amines and the comparative ability of two different amines to act as bases. The question also requires a definition of what a base is and an equation and mechanism for the reaction of another amine with ethanoyl chloride.

Step 2: Scan read for command words:

Term	Meaning
State	Express clearly
Describe	State the characteristics
Explain	Give a reason for
Give	Produce an answer from recall and/or the information in the question
Suggest	Use your knowledge and understanding to present a possible solution

Step 3: Apply this to the context of the question:

- *State the feature that accounts for the ability of an amine to act as a base.* It helps to recall the definition of a base.
- *Describe and explain the difference in the strength of ethylamine and salbutamol as bases.* This involves a comparison of the two compounds, relating to their strength as bases.
- *Give the equation for the reaction.* Use the reactants stated.
- *Suggest a mechanism for the reaction.* Apply what you know about the reaction of an amine with an acyl chloride to an unfamiliar compound.

Sample answer

(a) Amines contain a lone pair of electrons on the nitrogen atom, enabling them to act as bases or proton acceptors. ✓

(b) Ethylamine is a primary amine, whereas salbutamol is a secondary amine. ✓ Secondary amines are stronger bases ✓ because of the inductive effect of the groups attached to the N atom, which is stronger than that of the hydrogen atoms attached to the N in a primary amine. ✓ Salbutamol is a stronger base.

You should know

> What your textbook contains and how to use it for maximum benefit.
> The different styles of reading and when to apply each.
> How to use the specification to guide your reading.
> How to find, analyse and evaluate further reading resources.
> How to read an exam question and identify the relevant information.

Learning objectives

> To identify the different forms of writing and when to use each.
> To develop an effective form of note taking.
> To plan answers to unstructured questions.

Study skills

Writing is the means by which the knowledge and understanding you have developed throughout the course will ultimately be assessed. For this reason alone, it is one of the most important skills to develop.

If you progress beyond A-level chemistry, you may well be involved in the writing of papers and posters, but at this stage your experience of writing will be:

→ note taking
→ recording practical work
→ answering questions
→ report writing

At A-level chemistry you are not likely to encounter extensive writing activities, however, you do need to be able to provide a logical, well-structured piece of writing, which includes the use of correct terminology. In the exam you will have a variety of different question styles, some of which will be judged for QWC (quality of written communication). In this case the answers will be expected to be:

→ legible, with spelling, punctuation and grammar being accurate enough to make the meaning clear
→ appropriate for the purpose and subject matter
→ coherent and clearly organised, using correct subject-specific terminology

Students must make the meaning clear to be awarded the marks, so not only must words and phrases be correct, but also the sentences in which they are used must make sense.

Note taking

Note taking is largely for your own benefit and you may not be taught how to do this. Whichever method you adopt, it needs to work for you.

One method is the Cornell note-taking system (Figure 3.1), a system devised by Cornell University, where the page is laid out to provide space for adding annotations (usually in a margin) and a summary (usually at the foot of the page).

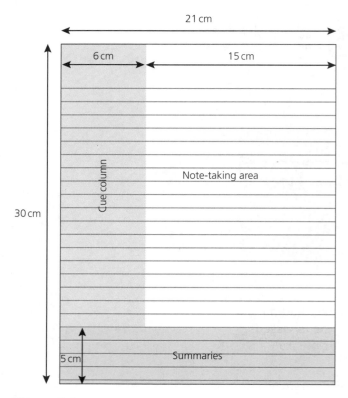

Figure 3.1

The idea behind this format is that, during the lesson, notes are made in the central part of the page, and that, after the lesson, these are summarised at the foot of the page and key points or questions (that are answered by the notes) are written in the margin. When you come to revise, you can then cover the central part of the page and use the questions as prompts to see what you remember. It is not just how the notes are set out that is important, but also how they are used. The Cornell method advocates reflecting on the content of the lesson soon after, and then reviewing all previous notes on a regular basis.

Your notes need to be well organised; they should have a title so that you know what they are about and that title should also set the context. You may wish to include the specification reference as well.

Key definitions need to be identified, as does other important information, such as practical procedures or equations. Underlining or highlighting this information can help it to stand out.

Calculations should be laid out so that each stage is shown clearly (see next page), including any equations in which values are substituted, and with the appropriate units. (More information on how to do this is included in chapter 1: Quantitative skills.)

Table 3.1

Bond	Bond energy/kJ mol^{-1}	Bond	Bond energy/kJ mol^{-1}
C–H	412	C–C	348
C=O	743	O–H	463
O=O	496		

The table below shows how (and how not) to lay out your calculation if asked to calculate the energy change during the combustion of propane, using the data given in Table 3.1 on the previous page.

The difference between...

A well laid out calculation		A poorly laid out calculation
$C_3H_8 + 5O_2 \rightarrow 3CO_2 + 4H_2O$		$C_3H_8 + 5O_2 \rightarrow 3CO_2 + 4H_2O$
Bonds broken:	Bonds made:	$8 \times 412 + 5 \times 496 + 2 \times 348 - 6 \times 743 + 8 \times 463$
C–H × 8	C=O × (2 × 3)	$= -1690\,kJ\,mol^{-1}$
$412\,kJ\,mol^{-1} \times 8 = 3296\,kJ\,mol^{-1}$	$743\,kJ\,mol^{-1} \times 6 = 4458\,kJ\,mol^{-1}$	
O=O × 5	O–H × (4 × 2)	
$496\,kJ\,mol^{-1} \times 5 = 2480\,kJ\,mol^{-1}$	$463\,kJ\,mol^{-1} \times 8 = 3704\,kJ\,mol^{-1}$	
C–C × 2		
$348\,kJ\,mol^{-1} \times 2 = 696\,kJ\,mol^{-1}$		
Sum of bonds broken:	Sum of bonds formed:	
$6472\,kJ\,mol^{-1}$	$8162\,kJ\,mol^{-1}$	
Overall energy change = bonds broken – bonds formed		
$6472\,kJ\,mol^{-1} - 8162\,kJ\,mol^{-1} = -1690\,kJ\,mol^{-1}$		

Summaries should be shortened versions of the key points of a topic, not a copy of what is in the textbook. A good way of getting used to doing this is to compare what is in a textbook with what is in a revision guide. The revision guide tends to just have the key facts and not the background or the examples. For example, for the topic 'Shapes of molecules', an OCR revision guide (ISBN 9781471842108) provides information as a summary table (Figure 3.2).

Number of bonded pairs of electrons	Number of lone pairs of electrons	Shape	Approximate bond angle	Symmetry
2	0	Linear	180°	Yes
3	0	Trigonal planar	120°	Yes
4	0	Tetrahedral	109.5°	Yes
5	0	Trigonal bipyramidal	90° and 120°	Yes
6	0	Octahedral	90°	Yes
3	1	Pyramidal	107°	No
2	2	Angular	104°	No

Figure 3.2 Shapes of molecules as covered in a revision guide

The OCR textbook (ISBN 9781471827068) describes each line of the table in detail, giving an example, a dot-and-cross diagram and an image of the shape (Figure 3.3).

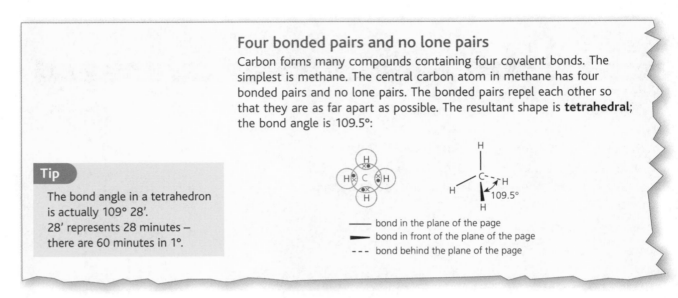

Four bonded pairs and no lone pairs

Carbon forms many compounds containing four covalent bonds. The simplest is methane. The central carbon atom in methane has four bonded pairs and no lone pairs. The bonded pairs repel each other so that they are as far apart as possible. The resultant shape is **tetrahedral**; the bond angle is 109.5°:

> **Tip**
>
> The bond angle in a tetrahedron is actually 109° 28'.
> 28' represents 28 minutes – there are 60 minutes in 1°.

—— bond in the plane of the page
■—— bond in front of the plane of the page
--- bond behind the plane of the page

Figure 3.3 Shapes of molecules as covered in a textbook

Recording practical work and its application to the exam

Throughout your course you will undertake practical work and will need to record the procedure followed and your results. You may be provided with a separate lab book to keep this work in. If you study chemistry beyond A-level, you will most certainly use a lab book and you will probably be expected to write up lab reports. The report should document the procedure in sufficient detail to enable someone else to carry out the same method and to achieve similar results.

The general format is to include a diagram (or list of equipment), a step-wise method (written in the third person, usually as a series of bullet points), a results table, some form of analysis and a conclusion. The exact content and format will depend upon the experiment being carried out. It may be that, for the majority of your work, you are provided with a method to follow. In this case, this should be secured into your notes and any amendments should be recorded alongside.

Writing a method

It is important that you can accurately describe how a practical is carried out because this is one of the skills that is now included in the written examination. Students often waste time listing apparatus, but this is not necessary because a labelled diagram or description of the method will cover this.

> **Activity**
>
> Choose a topic (it could be one you have already studied) and produce a one-page summary of your notes using the Cornell method described above. Make sure you include the cue questions and the summary of the main points.

Worked example 3.1

Writing in a logical sequence of steps

A student produced a sample of 1-bromobutane, starting from 28.1 g of butan-1-ol and an excess of powdered sodium bromide. Describe the procedure used and state any additional reagents required. Your answer should include a calculation of the minimum mass of sodium bromide, but no other quantities are required.

This is an example of the synthesis of an organic liquid. You need to describe the method in a logical sequence of steps. You are not asked to justify why you are carrying out particular steps, but you should refer to any safety precautions (beyond the obvious).

Start off by jotting down what the different stages of the procedure will be. It might be helpful to draw a rough sketch of the apparatus as a reminder.

(a) Mixing the reactants

(b) Reflux and then distillation of the product

(c) Purification of the product

(d) Drying the product

Having done this, you can then add a little more detail, again in note style. Name the reagents that will be used, how they will be prepared and any other equipment or resources that may be required. This may be dependent on the reagents involved in the process.

(a) Mixing the reactants: mass of solid NaBr; water to make solution; conc. sulfuric acid — dropping funnel and ice bath

(b) Reflux and then distillation of the product — organic liquid, so water bath or heating mantle; choice depends on boiling point

(c) Purification of the product: use of separating funnel and identification of organic layer; sodium hydrogen carbonate solution to remove acid (release pressure)

(d) Drying the product: add drying agent and filter (anhydrous calcium carbonate)

This then forms the plan for the answer you will write and following it enables you to produce a logical and well-structured response.

Activity

Write an answer to the question given at the start of Worked example 3.1, using the suggested plan as a guide.

Writing a conclusion

When you have generated some data — whether qualitative or quantitative — you will need to write a conclusion to summarise what your major findings were. This may also include what the errors were and how you could improve the practical.

Again, this is an aspect of the practical work that could be examined, and so writing in a logical and structured way is key to securing full credit.

 Exam tip

When you are faced with an unstructured question, such as describing a practical procedure, produce an outline plan. When you have written your response in full, put a single line through the plan to show that it was rough work. Proofread your work and amend any errors you spot.

Worked example 3.2

Evaluation of a practical

When a question asks you to *evaluate*, you are being asked to make a judgement about something — in this case the practical procedure that has been performed. You will need to use your own experience of the procedure as a reference point, comparing the information in the question with your knowledge and understanding. Read through the information given, including any results or other data, and annotate anything that stands out as being unusual.

A student carried out an experiment to determine the concentration of ethanoic acid, by titration against 20.0 cm³ aliquots of a 0.500 mol dm⁻³ solution of sodium carbonate. The results obtained are shown in Table 3.2.

Table 3.2

	Trial	1	2	3
Final burette reading/cm³	7.30	14.25	21.15	28.25
Initial burette reading/cm³	0.00	7.30	14.25	21.15
Volume of CH₃COOH used/cm³	7.30	6.95	6.90	7.10

Titres 1 and 2 are concordant; however, the titres seem rather low.

(a) When setting up the apparatus, the student failed to fill the burette correctly and left the gap between the tip and the tap full of air. State and explain the impact that this will have had on the results that the student recorded. (2)

The burette would deliver less solution than that indicated by the result.

(b) Suggest a way in which the student could gain more accurate results, using the same equipment and procedure. (2)

Using a bigger volume would give a smaller percentage error.

Step 1: Read through and annotate the question. Some examples have been added for you.

Step 2: Part (a) tests your knowledge of how a burette works and the fact that it is calibrated so that the volume delivered includes the amount of liquid in the portion of the burette beyond the tap. If this is not filled, then the actual titre would be greater, because when the tap is opened, the liquid will fill the gap before being delivered to the flask, so the total volume reading would increase.

Step 3: The question says *state and explain*, so you need to say *what* the impact is and give a *reason*.

Step 4: Part (b) asks you to *suggest* a way of increasing the accuracy using the same equipment and procedure. Looking at the results, the titres seem rather low and so one way to improve the accuracy would to decrease the impact of the equipment error — by making the titre larger.

For most measuring equipment, the manufacturer will provide the maximum error that is inherent in using that piece of equipment, which is usually ± half the smallest measurement that the equipment is capable of. This can then be used to calculate the percentage error:

$$\text{percentage error} = \frac{\text{equipment error}}{\text{measurement made with that piece of equipment}} \times 100$$

If you increase the size of the denominator, then the error becomes less.

To increase the size of the denominator, we need to increase the volume of acid that is added. However, this is linked to the number of moles that will react with the carbonate. Assuming the concentration of the acid remains the same, as this is what is being determined, and the moles of acid and the volume of the sodium carbonate are the same, then the only way to increase the number of moles is to increase the concentration of the carbonate:

$$n = \frac{v}{1000} \times C$$

Sample answer

(a) By leaving air in the burette between the tap and the tip, the student delivers less acid than suggested by the recorded titre. The difference is the volume that would fill the gap between the tap and the tip of the burette.

(b) Sodium carbonate reacts with ethanoic acid according to the equation:

$$Na_2CO_3 + 2CH_3COOH \rightarrow 2CH_3COONa + H_2O + CO_2$$

Two moles of acid react with every mole of sodium carbonate. The number of moles of sodium carbonate is determined using the volume and the concentration of the solution used:

$$n = \frac{v}{1000} \times C$$

To increase the volume of the acid, the concentration of which is fixed, the student needs to increase the number of moles that are reacting with the carbonate. If the aliquot of the carbonate remains at 20 cm^3, then the concentration of this must be increased to increase the number of moles present.

Short-answer questions

These can vary from a single word to a sentence or two, and are usually worth up to 3 marks. Some answers may be structured, where the question is broken into sections, each of which is answered in turn using spaces inserted into the question. Many of these questions will be based on knowledge, such as the recall of definitions or the identification of functional groups.

When starting an answer to a short-response question, do not repeat the question stem in the answer. This wastes time and space.

Writing answers to longer, unstructured questions

Some exam questions will have up to a page of blank space following the question, in which you are expected to present your answer. Many students allow themselves to be distracted by the amount of space available and end up spoiling an otherwise good answer by rambling in order to fill the space on the page. This is detrimental not just because it wastes time, but also because it can lose you marks if you go on to contradict yourself because you are waffling.

Longer-answer questions are worth more marks and usually cover more of the higher-order skills, such as evaluation, comparison, justification and analysis.

Activity

Look at one of the mark schemes for your awarding body and find the guidance for answering one of the level-of-response, unstructured questions. Use this information to complete the following table

Response level	Guidance on the communication of ideas
3	
2	
1	
0	

Although not all of the boards present this information in exactly the same way, or have identical response levels, they do all include this type of question, where marks are allocated for content and for how the information is presented.

Developing an appreciation of the hierarchical nature of explanation should help you structure your written answers and should ensure that you do not omit basic information. Students often lose marks because they fail to state what they think is obvious.

Worked example 3.3

Building an answer to a level-of-response question

Describe the mechanism for the reaction between but-1-ene and hydrogen iodide, and use this to name and explain the products formed.

This is a level-of-response-style question, so you need to get the chemistry correct *and* structure your answer.

The correct chemistry includes recognition that this is an electrophilic addition reaction and that, because of the position of the double bond, there are two possible products — 1-iodobutane and 2-iodobutane. The secondary carbocation is the more stable intermediate, so 2-iodobutane will be the major product. Marks will also be awarded for drawing the mechanism correctly, with all dipoles, curly arrows and charges shown and positioned accurately.

Sample answer A

The hydrogen iodide is attracted to butene and reacts to form iodoalkane, as shown in the diagram.

The answer has the correct dipole on the hydrogen iodide and forms a correct product. It states that the hydrogen halide is attracted to the butene. This would score just 1 mark because there is very little explanation and the mechanism is incomplete.

Typically, to score between 1 and 2 marks (the lowest level of credit-worthy response) would require an answer that got at least some of the chemistry right, including terminology. In reponses requiring mechanisms, most of the marks are awarded for the correct curly arrows and structures.

Sample answer B

The hydrogen iodide can form 1-iodobutane or 2-iodobutane depending on whether it adds to the first carbon atom in the alkene or the second one. The reaction is an electrophilic addition and occurs because the hydrogen halide is a polar molecule, which is due to the fact that iodine is more electronegative than hydrogen.

(a)

$$H_2C=C(H)-CH_2CH_3 \quad \rightarrow$$

$H^{\delta+}$
$I^{\delta-}$

(b)

$$H-CH_2-C^{+}(H)-CH_2CH_3$$

$+\ I^{\delta-}$

(c)

$$CH_3CHI\,CH_2CH_3$$

OR

$$CH_2ICH_2\,CH_2CH_3$$

This answer is better than answer A because it shows the carbocation intermediate produced after the addition of hydrogen, and it shows the second stage of the mechanism (although the charge on the iodine is wrong). Both products are correctly identified.

To score between 3 and 4 marks, more of the mechanism needs to be complete, the student needs to recognise that two products are formed and there needs to be reasoning evident in the answer, linking the key points.

Sample answer C

The electrophilic reaction between but-1-ene and hydrogen iodide produces two products, 1-bromobutane and 2-bromobutane. 2-bromobutane is the major product because the secondary carbocation intermediate that forms this product is more stable than the primary carbocation, which forms the minor product, 1-bromobutane.

$$H_2C=C(H)-CH_2CH_3 \rightarrow H-CH_2-C^{+}(H)-CH_2CH_3 \rightarrow$$

$H^{\delta+}$
$I^{\delta-}$

$:\!I^{-}$

$$CH_3CHICH_2CH_3$$
(major product as 2° carbocation is more stable)

$$CH_2ICH_2CH_2CH_3$$
(minor product as 1° carbocation is less stable)

This answer includes a complete and accurate mechanism and an explanation of the formation of the carbocation intermediates. It links the stability of these intermediates to the products formed and uses this to justify which will be the major product. This answer would be worth full marks.

To score between 5 and 6 marks, the mechanism needs to be complete and correct, and the major product identified.

The difference between...

A B-grade student	An A-grade student
• will identify what the question is asking and note the key points • will identify the majority of the key terms to be included • will use some of the key terminology correctly	• will identify what the question is asking and note the key points to include as bullet points. • will order the bullet points • will note any key vocabulary terms that should be included • will use this plan to write an answer • will tick the points off in the question as each is addressed in the answer • will cross through the plan once the answer has been drafted

Exam tip

When writing a longer, unstructured answer, first concentrate on including as much correct chemistry as possible. Then focus on the clarity and coherence of the answer and drafting it in a logical progression.

Not all long-answer questions will be marked as level-of-response questions, but it is useful to have a plan and to structure your answers so that you include all the relevant information. Keeping it in a logical sequence means that you are less likely to omit information.

Using terminology correctly

At A-level, students are expected to be able to use the correct terms for describing particles and processes. Students often lose marks by careless use of terminology. As an A-grade student you need to be pedantic about your use of language from the start of the course; think carefully about the words you choose and be specific when writing descriptively.

One of the ways to help you in the correct use of terminology is to identify the key terms in the question and the context in which they are used.

Worked example 3.4

Using terminology

Nitrogen monoxide is present in the exhaust gases of aircraft. NO radicals act as homogeneous catalysts by breaking down ozone in the stratosphere.

Explain — using equations to illustrate your answer — the meaning of the terms homogeneous and catalyst as applied to this mechanism.

Before answering the question it is important to consider the meaning of the highlighted terms:
● Radical — a species with an unpaired electron (which is shown as a dot when writing the formula), and which is reactive.
● Homogeneous — the catalyst and the reactants are in the same state.
● Catalysts — speed up reactions by offering an alternative route with a lower activation energy, and are regenerated during the reaction.

Step 1: Write the equations that describe this process — remembering to designate the NO as a radical:

$$\cdot NO(g) + O_3(g) \rightarrow \cdot NO_2(g) + O_2(g)$$

$$\cdot NO_2(g) + O(g) \rightarrow \cdot NO(g) + O_2(g)$$

Step 2: Explain the meaning of the term homogeneous in the context of the question. Using state symbols in the equation, you can now illustrate that the catalyst is homogeneous because it is in the same (gaseous) state as the reactants (and products).

Step 3: You need to refer to the definition of the term catalyst, but in the context of this question. The equation shows that the NO radical is used and regenerated, so *this* is the meaning you need to use here. Referring to a catalyst as providing an alternative route with a lower activation energy is irrelevant and so would not score any marks.

✅ Exam tip

Learn definitions and recall these when answering a question. Select the part of a definition that is relevant to answering the question.

✅ Exam tip

Avoid anthropomorphism — the attribution of human characteristics to chemical species — in your answers. Do not refer to atoms as though they have thoughts or feelings. They do not want to gain a complete outer shell of electrons; it happens because this is the most stable arrangement.

Application to the exam

Shorter-response questions

Worked example 2.3

Maintaining the pH of the blood is critical for the healthy functioning of the body and is achieved by the buffer system involving carbonic acid (H_2CO_3) and the carbonate ion (HCO_3^-). Explain, with the aid of an equation, how this buffer solution works when acid is added to the blood. (4)

How to approach this question

Although this question is not set out as separate steps, these are evident in the question — you need to explain how the *blood buffer system* works when acid is added, and you need to include an equation.

Identify what the question is about and highlight any information that you can use in your answer. The two species highlighted will be needed for writing your equation. Failing to include this equation will mean that you cannot score full marks. Allow yourself 6 minutes to write an answer to this question.

Sample answer

When H^+ ions are added to the blood, the equilibrium, $H_2CO_3 \rightleftharpoons HCO_3^- + H^+$ ✓, shifts to the left ✓ because the HCO_3^- ions that are present in excess ✓ react with the added H^+ ions, reforming the acid and restoring the equilibrium ✓.

There is no need to include that the H^+ ions have been produced during exercise — the question is about how the buffer works. It is important to recognise that it is the excess of hydrogen carbonate ions that enables the buffer to react with, and remove, the hydrogen ions.

Longer, unstructured responses

Worked example 2.4

The equation for the decomposition of sulfur dichloride dioxide is:

$$SO_2Cl_2(g) \rightarrow SO_2(g) + Cl_2(g)$$

The reaction rate can be followed by measuring the pressure of the gases in the reaction vessel and using this to calculate the concentration of SO_2 at given times.

A graph of the concentration against time for this reaction is shown in Figure 3.4.

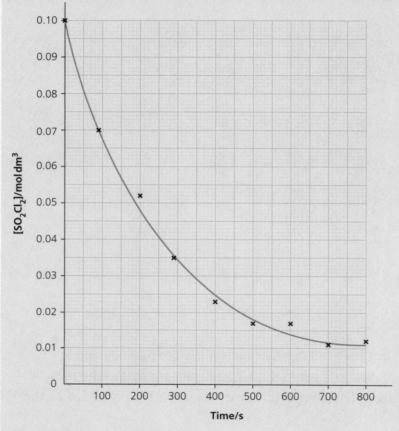

Figure 3.4

From the graph determine:

(a) the initial rate of the reaction

(b) the rate constant for the reaction

Your answer must show full working, using the graph as appropriate.   **(6)**

How to approach this question

There are three components to this question — the determination of the initial rate, the determination of the order of the reaction and the calculation of the rate constant.

The graph is a concentration–time graph. The shape of the graph can be used to predict the order of the reaction. The order can also be determined by finding the half-life of the reactant — the time taken for the concentration of that reactant to halve.

● The initial rate is found by taking the gradient at the start of the reaction.

● The order of the reaction is found using half-lives.

● The rate constant is then found by substituting numbers into the correct rate equation, which can be written once you know the order of the reaction.

The question clearly states that the graph should be used in the answer and that full working must be shown.

Sample answer

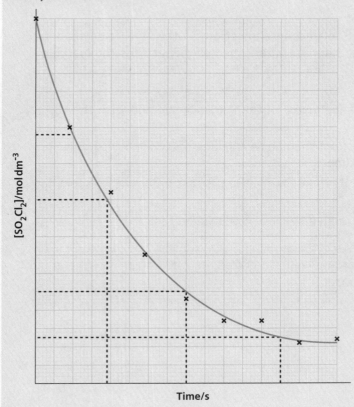

(a) The initial rate of the reaction $\Delta y/\Delta x$ at time 0 is:

$$\frac{0.100 - 0.068\, mol\, dm^{-3}}{100 - 0\, s}$$

$$= \frac{0.032}{100}\, \frac{mol\, dm^{-3}}{s} \quad \checkmark$$

$$= 3.2 \times 10^{-4}\, mol\, dm^{-3}\, s^{-1} \quad \checkmark$$

(b) Half-life: $0.100 - 0.050\, mol\, dm^{-3} = 190\, s$

Half-life: $0.050 - 0.0205\, mol\, dm^{-3} = 400 - 190\, s$ or $210\, s$ $\checkmark$

These are roughly the same, so the reaction is first order with respect to SO_2Cl_2. $\checkmark$

$$rate = k[SO_2Cl_2]$$

So:

$$k = \frac{rate}{[SO_2Cl_2]} \quad \checkmark$$

$$k = \frac{3.2 \times 10^{-4}\, mol\, dm^{-3}\, s^{-1}}{0.1\, mol\, dm^{-3})}$$

$$= 3.2 \times 10^{-3}\, s^{-1} \quad \checkmark$$

This answer would score full marks. This is a level-of-response question, so the level is decided first and then the mark. The following can be used to help see where the answer satisfies the marking criteria:

● Tangent drawn at $t = 0$.
● Calculation of the initial rate is within the range given on the mark scheme and with the correct units.
● Graph shows evidence of drawing lines down from the concentrations of $0.05\, mol\, dm^{-3}$ and $0.025\, mol\, dm^{-3}$, showing the corresponding times.
● States that because the half-lives are constant, then the reaction is first order with respect to SO_2Cl_2.
● The correct rate equation is written and rearranged to make k the subject.
● Correct calculation and units of k.

Throughout the question, communication is clear and well structured, and it is clear how the graph has been used. The calculations and units are correct.

You should know

> The importance of using the correct scientific terminology throughout your answers.

> How to structure a plan for an unstructured answer, so that your response is comprehensive and logical.

> The importance of proofreading your answers to ensure they make sense, and to tick off the points in questions as you answer them, so that you are not leaving anything out.

4 ▲ Practical skills

Learning objectives

> To develop the skills to help you plan for and understand practical techniques and activities.
> To develop the skills required to pass the practical endorsement.
> To be able to explain the underlying scientific principles in a practical method.
> To evaluate a method and suggest improvements.
> To apply the above to the answering of questions based on practical techniques.

Study skills

Practical work is a key component of A-level chemistry and all students, regardless of which specification they are following, are required to demonstrate competency for the Common Practical Assessment Criteria (CPAC), which are broadly divided into five sections:

→ Following written procedures
→ Applying investigative approaches to practical work
→ Safely using equipment and chemicals
→ Making and recording appropriate observations
→ Researching, referencing and reporting

If you meet the assessment criteria, and the majority of students are expected to do this, then you will be awarded a pass on your A-level practical endorsement certificate. If you are going on to study a scientific discipline at university, it is likely that you will need to have passed the practical endorsement. This may even be a requirement for some other, seemingly unrelated courses. For example, to study the finance and accounting BSc Honours course at the University of Bath, you will be expected to pass any separate science practical endorsement if you are taking GCE A-level in a science subject. Always check out current university entrance requirements.

The number and type of practicals carried out will vary between the exam boards and from school to school, and not all practical work will be linked to the practical endorsement. Your school may instruct you to record practical work in a lab book. This provides an excellent first experience of recording scientific information — particularly useful for those thinking of studying science at university.

The difference between...

A B-grade lab book	An A-grade lab book will also have the following
• Includes dates and titles for all work. • Methods included are written in such a way that someone else could follow them, or instruction worksheets are included. • Results are neatly and accurately tabulated. • Sources are referenced.	• An indication of where the experiment fits within the curriculum, e.g. rate experiments would be labelled kinetics; preparation for aspirin as organic synthesis. • Modifications to methods — such as changes in volumes of reagents used or the time taken for a particular stage — are included as annotations to instructions. • Results include a record of equations used and how the data are manipulated. • Sources are referenced, including the date of access for web resources.

Practical skills will be assessed in the written exams and, for WJEC and CCEA, in the laboratory through a practical examination. Some exam boards assess the practicals across all papers; others on particular ones.

The practical skills that are assessed through exam questions account for a minimum of 15% of the overall marks. In order to be able to answer these questions, you need to be competent in the skills shown in Table 4.1.

Activity

Check your exam board specification to determine how they assess practical skills. Find out on which papers this occurs and, where possible, how many marks are allocated.

Table 4.1 Practical skills required for A-level chemistry

Knowledge and understanding (AO1)	• The ability to select and describe practical procedures.
Independent thinking (AO2)	• The ability to solve problems set in a practical context.
Application (AO2)	• The ability to apply scientific knowledge to a practical context. • The ability to use and present data appropriately.
Analysis (AO3)	• The ability to draw conclusions from data. • The ability to select, process and analyse data using appropriate mathematical skills, including plotting and interpreting graphs.
Evaluation (AO3)	• The ability to consider margin of error, accuracy and precision of data. • The ability to evaluate a scientific procedure and suggest improvements. • The ability to ability to evaluate results and draw conclusions based on them.

This chapter aims to support you in the development of your practical skills and in understanding how they will be assessed in the exam. It will encourage you to think about the purpose of each practical — not only how to carry it out safely and effectively, but the underlying chemistry, what to do with the results, the limitations and how the results and the procedure could be improved.

Activity

Table 4.2 gives examples of some of the different practical activities you might be required to complete. See if you can match the apparatus and techniques to the different activities, using the example given for guidance and by consulting the specification for your exam board for more information. Decide whether the data generated will be qualitative or quantitative, and add this to the table.

Table 4.2 Analysing the different practical activities

Practical skill	Apparatus required or suggested reagents	Technique(s) used	Application	Qualitative or quantitative technique?
Make up a volumetric solution and carry out a simple acid–base titration	Pipette, burette, volumetric flask	Titration	Acid–base titration for the determination of the concentration of a solution	Quantitative
Identification of unknown inorganic ions				
Identification of organic compounds or functional groups				
Measurement of enthalpy change				
Investigating rate of reaction				
Synthesis of an organic compound				

Practical skill	Apparatus required or suggested reagents	Technique(s) used	Application	Qualitative or quantitative technique?
Measurement of pH changes				
Investigating electrochemical cells				

Determining your practical method

Preparing for the practical

At A-level, you should be clear about the purpose of each experiment that you carry out and where it fits into the curriculum. This is because practical work can be used in a variety of different ways, from introducing a concept, to identifying a particular functional group in a molecule, to determining the rate equation for a particular reaction, the concentration of a particular reactant or some other variable. Practical work can also be used as part of an open investigation, where a hypothesis is proposed and then suitable practical procedures determined and employed in order to try to prove the hypothesis.

Before undertaking any practical work it is important to read through the instruction sheet or watch the demonstration of the practical technique. Make sure that you know the order in which you need to complete each step, what equipment you need to use and how you will set it up, and any contingencies (such as the need to cool a solution prior to use). You may also need to consider the volumes of reagents that would be suitable to use and whether you need to make up any solutions or to dilute any that you have been provided with. If this is the case, it is useful to record the volumes of the solution and water that you will need to use.

You need to be able to follow a written procedure to meet the criteria for CPAC 1. You also need to be able to follow instructions for the assembly of apparatus, whether from written instructions or a diagram. The written exam may require you to describe a procedure; to write or complete a method or to suggest improvements to a given method. In order to help you visualise how you have carried out practical work when you come to the exam, you need to maintain your focus when you are carrying out lab work. Think about why you set up the equipment in the way you do, because this will help you to understand the implications on your results if it is not set up correctly.

> **! Common pitfall**
>
> Students often miss out on marks by drawing apparatus incorrectly. Make sure you can draw how the equipment should be assembled for all of the required practical activities for your exam board.

The difference between...

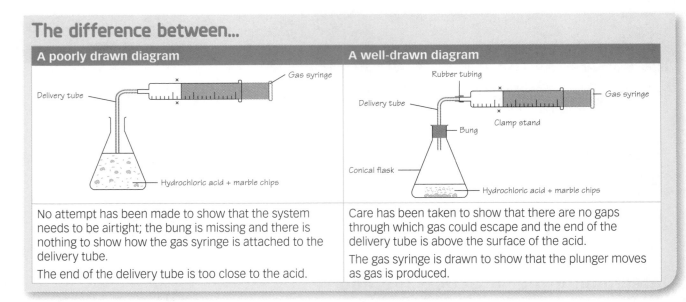

A poorly drawn diagram	A well-drawn diagram
No attempt has been made to show that the system needs to be airtight; the bung is missing and there is nothing to show how the gas syringe is attached to the delivery tube. The end of the delivery tube is too close to the acid.	Care has been taken to show that there are no gaps through which gas could escape and the end of the delivery tube is above the surface of the acid. The gas syringe is drawn to show that the plunger moves as gas is produced.

Use the points in the checklist in Table 4.3 to ensure that you have drawn your diagrams correctly and to avoid common errors.

Table 4.3 Checklist for drawing apparatus

Point	Example and explanation
Is the apparatus set up safely?	A naked flame should not be used to heat a flammable liquid; stoppers should not be used where pressure can build up.
Is the apparatus assembled correctly?	There should be no leaks for equipment designed to collect gases; water should go into a condenser at the bottom inlet.
Is it drawn correctly?	The diagram should be a 2D section through the equipment; clamps are signified by an X either side of the apparatus being clamped and a Bunsen burner is shown as an upward arrow; there should be no breaks in outlines except where openings are indicated.

Activity

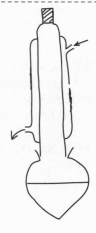

Figure 4.1

A student has drawn the diagram in Figure 4.1 to represent the apparatus used to prepare an ester, by heating an alcohol with an organic acid, under reflux. Highlight the errors on the student's drawing and indicate what they should do to correct them.

As an A-grade student you will apply your understanding of the procedure being carried out, and how the equipment works, to the correct drawing of apparatus.

Risk assessment

All procedures that you follow will have had risk assessments carried out by supervising teachers, so you should be aware of the main hazards and how to prevent them. This includes recognising and understanding the hazard symbols used on reagent bottles, as well as having an appreciation of when and why reagents should be used in a fume cupboard and the risks associated with particular pieces of equipment or techniques. You should also be familiar with Hazcards and how to use them to find out information about chemicals.

Whilst it is less likely that an exam question would ask explicitly about safety, the boards that have practical assessments allocate some marks to working safely, and it is also monitored in any of the required practical activities that you will undertake. The checklist for drawing diagrams (Table 4.3) includes guidance on how to consider safety when setting up equipment.

You should also be aware of the hazards associated with using flammable, corrosive or toxic chemicals, as well as concentrated solutions of acids and alkalis. You also need to know how to take precautions to protect yourself, over and above standard lab safety.

Sequencing the steps of the procedure

It is essential that you understand the practical procedures you carry out in order to be able to describe them accurately. Examiners' reports highlight the need for students to organise their thoughts into a logical sequence, particularly for extended-response questions; undertaking a variety of practical activities helps develop this skill.

> 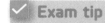 **Exam tip**
>
> As always, read the question twice, ensuring that you pay particular attention to any information about chemicals and procedures. Then answer each part of the question in turn.

Worked example 4.1

Describing a method in a logical sequence of steps

Propan-1-ol can be oxidised under carefully controlled conditions to make propanal, a low boiling point liquid.
(a) Describe how a student could prepare a sample of an aldehyde from 25 cm³ of the alcohol. Include a labelled diagram and details of the reagents required.
(b) Describe how the student could confirm experimentally that the product made is an aldehyde.

Step 1: First decide what type of alcohol propan-1-ol is. It is a primary alcohol, which can be oxidised to propanal and could be further oxidised to propanoic acid. In order to remove the propanal, which has a lower boiling point than propanoic acid, the apparatus needs to be set up for distillation, rather than reflux. ➡

Step 2: Draw a diagram similar to that shown in Figure 4.2, labelling the reactant and the product. The question states that propanal is a liquid with a low boiling point, so cooling the distillate using iced water is recommended.

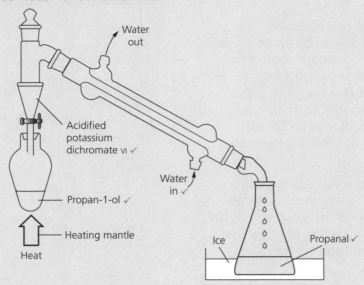

Figure 4.2 Apparatus used to distil propanal after oxidation of propan-1-ol

Step 3: The answer should recognise that the reagents are placed in the flask and heated and that the distillate is collected and cooled. These points are all evident from the diagram of the apparatus, so only a brief description of the procedure is required. For example:

Place the propanal and acidified potassium dichromate(VI) in the flask and heat gently, collecting the distillate in a conical flask, cooled in iced water. The acidified potassium dichromate(VI) will change colour from orange to green.

Step 4: Part (b) of the question is asking you for a qualitative test for an aldehyde. Either of those in Table 4.4 would be acceptable.

Your answer should include the reagent used plus the positive result with that reagent. For example:

Table 4.4 Test for an aldehyde

Tollens' test	Fehling's test
Add Tollens' reagent — an alkaline solution of ammoniacal silver nitrate — to the aldehyde and warm in a water bath.	Add Fehling's solution (a copper(II) salt) and warm in a water bath.
If an aldehyde is present, then a silver mirror forms. ✓	A red precipitate (copper(I)) forms if an aldehyde is present. ✓

The student places 1 cm³ of the distillate into a clean test tube and adds Tollens' reagent (ammoniacal silver nitrate) before placing the test tube into a beaker of hot water. The presence of a silver mirror coating the test tube confirms that propanal has been made.

Step 5: Go back through the question and tick off the parts answered.

Propan-1-ol can be oxidised under carefully controlled conditions to make propanal, a low boiling point liquid.
(a) Describe how a student could prepare a sample of an aldehyde from 25 cm³ of the alcohol. Include a labelled diagram and details of the reagents required.
(b) Describe how the student could confirm experimentally that the product made is an aldehyde.

The difference between...

See Figure 4.2.

A B-grade answer	An A-grade answer
• Identifies that reactants are heated but does not specify the heating source used. • The top of the flask containing the reactants is stoppered to prevent the vapour from escaping. • The oxidising agent is mixed with the propan-1-ol in the pear-shaped flask.	• Heating source used is a heating mantle or water bath (because the reactants are flammable). • A thermometer is indicated in the flask containing the reactants, to record the boiling point of the distillate. The bulb of the thermometer is in line with the point where the vapour enters the condenser. • The oxidising agent is added dropwise to the propan-1-ol in the flask. • Anti-bumping beads are indicated in the flask being heated.

Apply the underlying scientific principles

Application of knowledge

The application of knowledge and understanding (AO2) accounts for up to 44% of the total A-level mark. To achieve these marks students are expected to be able to explain the science underlying practical activities. It is therefore important to think about what is happening in each and every practical activity, and why. In order to do this, you need to know where the practical fits into your studies, in terms of the subject matter and the skills assessed.

Whenever you carry out a procedure, you should have an idea of what that procedure will achieve and, whenever you answer a question, you need to classify that question in terms of the topic it fits into, whether that be kinetics, synthesis, neutralisation or analysis. Once you have done this you will be much better placed, not only to describe what is happening, but also to say *why* it is happening. For example, when choosing an indicator for a reaction, you need to know what an indicator is, how it works and then to look at the particular reactants for the given reaction.

Activity

Sketch graphs to show the titration curves obtained when the titration is between:

(a) a strong acid and a strong alkali
(b) a weak acid and a strong alkali
(c) a strong acid and a weak alkali
(d) a weak acid and a weak alkali

Using Table 4.5 suggest which indicator would be an appropriate choice for each titration.

Table 4.5

Indicator	pH range	Suitable for which titration(s)
Phenolphthalein	8.2–10.0	
Phenol red	6.8–8.4	
Bromocresol green	3.8–5.4	
Methyl orange	3.2–4.4	

Worked example 4.2

Demonstrating knowledge of the underlying concept

A student is preparing a sample of benzoic acid by the hydrolysis of the ester methyl benzoate, according to the flow diagram shown in Figure 4.3.

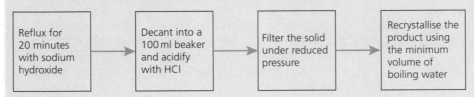

Figure 4.3

(a) What is the purpose of acidifying with the hydrochloric acid in stage 2?

(b) Describe the steps involved in the recrystallisation of the acid from the solvent (water), stating what property of benzoic acid this process depends on.

Step 1: Look carefully at each stage of the process shown in the flow diagram. The hydrolysis is being carried out using sodium hydroxide, which is an alkali.

The purpose of the hydrochloric acid will be to neutralise any unreacted sodium hydroxide from the reaction mixture.

Step 2: This is where you need to organise your answer into sequential steps:

Dissolve the product in the minimum amount of hot water.

(Stating *minimum* volume shows that you understand that you need to just get all of the solute to dissolve, so that the maximum amount recrystallises as the solution cools.)

Cool the solution to produce crystals and wash with cold solvent.

The question asks you to *describe*, so you would not need to include the information in brackets to be awarded the mark; however, should the question ask you to *explain* the process, you will need to understand and write about why each stage is carried out.

Step 3: Now relate the procedure to the property of benzoic acid that it depends upon:

Benzoic acid is more soluble in the hot solvent.

In an exam, you should use all the information given in the description of a practical procedure to help you answer the question. In addition to reading the question twice, it helps to underline or highlight the key points of the question. This could include: equations or quantities that you may need for calculations; stages in the procedure that you do not want to overlook; any guidance on how to present your answer.

Worked example 4.3

Highlighting the information in the question

A student carried out an experiment to find the enthalpy of reaction for the following reaction:

$$Na_2CO_3(s) + 2HCl(aq) \rightarrow 2NaCl(aq) + CO_2(g) + H_2O(l)$$

The student measured 50 cm³ of dilute hydrochloric acid into a polystyrene cup and added a sample of sodium carbonate from the weighing bottle.

The results were recorded in Table 4.6.

Table 4.6

Mass of weighing bottle and carbonate/g	57.86
Mass of weighing bottle after adding carbonate/g	54.02
Temperature of acid at start/°C	22.0
Temperature of acid after adding carbonate/°C	28.5

Calculate the enthalpy change for the reaction in kJ mol⁻¹.

Step 1: Read through this question and underline the volume of acid (*50 cm³*) and the fact that the carbonate is added from a *weighing bottle*. Also underline the units that the answer should be given in (*kJ mol⁻¹*).

Step 2: Look at the table, which enables the mass of carbonate used to be found, along with the temperature change for the reaction. Using the data from the question:
- Calculate the mass of carbonate used:

    $57.86 - 54.02 = 3.84$ g

- Calculate the change in temperature (the difference between the end and start temperatures of the acid):

    $28.5 - 22.0 = 6.5$ °C

Step 3: Write out the equation for the energy given out:

$$Q = mc\Delta t$$

where *m* is the mass of the solution (acid) and *c* the specific heat capacity of water, which is given on the data sheet. (The acid used is dilute and so using the specific heat capacity of water is appropriate here.)

Substitute values into this equation:

$$Q = 50 \text{ (the volume of acid)} \times 4.18 \text{ J g}^{-1}\text{K}^{-1} \times 6.5 \text{°C}$$

$$Q = 1358.5 \text{ J}$$

Step 4: Calculate the number of moles of sodium carbonate, by dividing the mass used by its relative formula mass (Na_2CO_3 $(23.0 \times 2) + 12.0 + (16.0 \times 3) = 106.0$):

$$\text{moles} = \frac{\text{mass}}{M_r}$$
$$= \frac{3.84}{106.0}$$
$$= 0.036 \text{ mol}$$

Step 5: You can now work out the enthalpy change for the reaction as follows:

$$\text{enthalpy change} = \frac{Q}{\text{moles}}$$

(Remember that when heat is given out you use a minus sign.)

$$\text{enthalpy change} = \frac{-1358.5}{0.036} = -37736.11 \text{J mol}^{-1}$$

Step 6: The question states that the answer should be given in kJ mol^{-1}, so you have to convert J to kJ by dividing by 1000, giving -37.74 kJ mol^{-1}.

Unless the question specifies the number of decimal places or significant figures, you should quote the answer to the same number as the least accurate value in the question, in this case the volume, and so your answer would be -38 kJ mol^{-1}.

Recording data

Once you are clear about how you are going to carry out an experiment safely, you then need to think about how you will record the data generated. This covers CPAC 4 — making and recording appropriate observations. The method of recording will depend on the type of data that are generated and whether you are recording measurements (quantitative data) or observations (qualitative data).

The difference between...

Quantitative data	Qualitative data
Any data where measurements produced by an externally validated scale are involved.	Any data where observations or descriptions are recorded and where there is no spectrum of results.
They should include appropriate units.	No units are needed.

Recording qualitative data

While much of the data generated by your practical work will be quantitative, the identification of particular ions or functional groups is qualitative. You are expected to be able to describe the tests for given anions (such as carbonates, sulfates and halides), cations (such as metal or ammonium ions) and functional groups (including COOH, C=C, aldehydes, ketones and alcohols), which are included in your exam board's specification. You should also be able to write ionic equations to represent the reactions.

As an A-grade student you will also be able to suggest an appropriate sequence for carrying out tests, depending on which species you are testing for and/or which reagents you have been given.

Activity

Qualitative analyses are based on observations. These can include colour changes, formation of a precipitate, the miscibility of different solutions and the production of a gas.

Complete Table 4.7 to give examples of each.

Table 4.7

Observation	What it indicates	Example	
Bubbles/effervescence	A gas is produced	Reaction of a carboxylic acid with a carbonate	
Colour change			
Precipitate forms			
Two layers appear			

The difference between...

It is important to give sufficient description in observations to secure the marks. Below are two possible responses to the following question:

Name a suitable reagent and state the observation obtained to confirm that a synthesis has produced a carboxylic acid. (2)

A B-grade answer	An A-grade answer
Add calcium carbonate ✓, which will react to give carbon dioxide.	Add a solution of sodium hydrogen carbonate ✓ to the carboxylic acid. Bubbles of gas would be observed. Bubbling the gas through limewater would enable the gas to be identified as carbon dioxide, if the solution went cloudy. ✓

Explaining observations

Many questions about qualitative procedures will ask students to describe their observations. Again, attention to detail is important: bubbles or effervescence tell you that a gas is being given off; a solution turning cloudy is due to the formation of a precipitate — it is important to state that a precipitate forms, rather than using the word cloudy.

Sometimes colour changes are not as clear-cut as expected. As an A-grade student you should be able to suggest why. Take the example of the oxidation of an alcohol by acidified dichromate(VI): the resultant solution is often murky brown, which you should recognise as being due to the mixing of the orange Cr^{6+} ions with the green Cr^{3+} ions, because not all alcohol molecules are oxidised at the same time and there may be an excess of a particular reactant.

Worked example 4.4

Interpreting qualitative data

A student was given four colourless solutions that were, in no particular order: hydrochloric acid, lithium chloride, magnesium sulfate and sodium carbonate.

(a) Barium nitrate was added to each of the solutions in turn. A white precipitate was formed in one of the test tubes. Suggest which solution was present in this test tube and write an ionic equation for the reaction taking place.

(b) When dilute nitric acid and silver nitrate were added, two of the solutions produced identical results. State what this result was and name the solutions that produced this result.

(c) Suggest a method of distinguishing between the two solutions that gave the same result when treated with silver nitrate.

Step 1: (a) Barium ions react with a sulfate to form a white precipitate of barium sulfate. So the solution in the test tube must be magnesium sulfate.

Step 2: Write the ionic equation:

$$Ba^{2+}(aq) + SO_4^{2-}(aq) \rightarrow BaSO_4(s)$$

Remember to show state symbols and just the ions that are forming the precipitate. (These are the ones that take part in the reaction.)

Step 3: (b) Again identify what the reagent (silver nitrate) is used to test for — the presence of halide ions — and then look for this within the given solutions. Chloride ions will be present in both hydrochloric acid and lithium chloride solutions, so these will both react with silver nitrate to form a white precipitate.

Step 4: (c) Consider the two solutions — hydrochloric acid and lithium chloride — and identify all the ions present: H^+ and Cl^- or Li^+ and Cl^-. As both solutions contain chloride ions and these were detected using the silver nitrate, you need a way to distinguish between hydrogen ions and lithium ions. You should recall that a carbonate reacts with dilute acid to produce bubbles of gas; alternatively, you could choose to carry out a flame test to identify the lithium ions in the lithium chloride solution.

Remember to state the positive result for each test you choose to carry out.

Recording quantitative data

Quantitative data are those that are measured. The recording of these data will need to include units and, very probably, repeats.

Table 4.8 is an example of how data are often recorded.

Table 4.8 Volume of gas produced during the reaction of an acid with a carbonate

Time/s	0	60	120	180	240	300
Volume of CO_2/cm^3	0	40	71	96	114	114

Readings within the table should be recorded to a consistent number of significant figures, determined by the resolution of the device being used to measure the data or the uncertainty in the measurement.

! Common pitfall

Students often miss out on marks when they fail to include units in their tables. Results should be recorded within a table that is clear and easy to read and with units given alongside headings.

Activity

Suppose you are going to carry out a practical to compare the enthalpies of combustion of three different alcohols. You are planning to use spirit burners, each one containing a mass of one of the three different alcohols, which you will burn and use to heat a given volume of water contained in a boiling tube clamped above the spirit burner. Draw a suitable table to record the results of this experiment.

Knowing how you will calculate the enthalpy of combustion (and which equation you will use) will help you to decide on the measurements you will need to take.

When recording results, these should reflect the precision of the instrument used in the measurement. As an A-grade student, you will understand the difference between accuracy and precision and use each term in the correct context.

Accuracy reflects how close a result is to the true value (of a measurement), whereas precision denotes the closeness of agreement between repeated measurements. For example, a burette reading could be given as $12.60\,cm^3$ or $12.65\,cm^3$ but no other values in between, because the smallest division of the scale is $0.10\,cm^3$ and so the value is determined as lying either on or between the gradations.

Accuracy should be limited to either the precision of the data (given in the question) or the accuracy of the apparatus. This should be remembered, particularly when any further processing of the data takes place.

Precision refers to how close repeated results are and is used in determining the values to include in mean calculations.

Take it further

More information on the correct use of terminology can be found in *The Language of Measurement*, published by the ASE (ISBN 978 0 86357 424 5). Exam boards may also have guidance on their websites so you should check that you are familiar with this.

Using the data generated by the practical

Data generated in a practical may need further processing using a variety of mathematical techniques. These include:

→ subtraction — when working out temperature changes for enthalpy experiments

→ mean calculation — when selecting concordant titres to find the mean in neutralisation reactions

→ taking logs or finding reciprocals — calculating $1/t$ for initial rate experiments

(Columns to carry these out may be included in the original table or you might prefer to produce a separate table.)

→ drawing of graphs and determining gradients from those graphs (further information on how to do this is covered in chapter 1 (quantitative skills))

! Common pitfall

Students lose marks when they cannot correctly identify the independent variable (the values of which are chosen) and the dependent variable (the values of which are measured), or when they do not put the independent variable in the first column of a table and the dependent variable(s) in subsequent ones.

Exam tip

Remember that the independent variable is always plotted on the x-axis of a graph and the dependent one on the y-axis, and that these axes should be labelled and include the correct units.

Activity

Carrying out several full experiments to provide enough data to determine the initial rate of reaction is time consuming, so sometimes an approximation is used. Clock reactions are ideal for this because they allow the measurement of time taken for a certain amount of product to be made, indicated, for example, by a change in colour or the formation of a precipitate. The rate to this point is taken to be proportional to $1/t$ (1/time) for each reaction, and the reaction is repeated with varying concentrations of the reactants.

The reaction between hydrogen peroxide and iodide ions in acidic solution can be set up as such a clock reaction:

$$H_2O_2(aq) + 2H^+(aq) + 2I^-(aq) \rightarrow I_2(aq) + 2H_2O(l)$$

A small (known) quantity of thiosulfate ions is added to the reaction mixture. The thiosulfate reacts with the iodine produced, turning it back into iodide ions:

$$2S_2O_3^{2-}(aq) + I_2(aq) \rightarrow S_4O_6^{2-}(aq) + 2I^-(aq)$$

Starch solution is added to the reaction mixture because it turns blue in the presence of iodine and acts as an indicator.

The results obtained in such an experiment are shown in Table 4.9.

Table 4.9

[KI]/mol dm^{-3}	[H$_2$O$_2$]/mol dm^{-3}	Time/s	1/t
0.025	0.025	55	0.02
0.025	0.050	25	
0.050	0.025	26	
0.050	0.050	12	

(a) Complete the table.

(b) Deduce the order of the reaction with respect to each reactant.

(c) Write a rate equation for the reaction.

(d) Explain, in terms of the reactant particles, why there is a delay in the appearance of the blue-black colour.

(e) Suggest how long it would take for the solution to turn blue if the concentration of potassium iodide was 0.025 mol dm^{-3} and the concentration of hydrogen peroxide was 0.075 mol dm^{-3}.

Evaluating the method

Assessment objective 3 requires students to be able to evaluate information, make judgements, and to develop and refine practicals. This includes the treatment of errors as well as an appreciation of the limitations of methods and equipment. All students should be able to select the correct results to process from any given set, and should also be able to state what impact given errors in the procedure will have on a particular result.

Equipment errors

You need to understand the limitations of the equipment being used to take a particular measurement. Each piece of equipment has a limit to its precision. When using a balance that weighs to two decimal places, you may see the value of the second decimal place move up and down a little because the balance has an inbuilt error. The error range is usually ± half the smallest measurement possible, so in this case ± 0.005 g.

Manufacturers usually give details of the maximum error for each piece of equipment. Remember that the percentage error for a piece of equipment is calculated as follows:

$$\text{percentage error} = \frac{\text{uncertainty of instrument}}{\text{quantity measured with instrument}} \times 100$$

It is also important to remember that if the equipment is used twice to calculate a value (for example, reading the start and finish values from a burette to work out the titre), then the maximum error is doubled). For example, a burette has an uncertainty of ± 0.05 cm³. If the mean titre for a neutralisation was 21.45 cm³, calculate the percentage uncertainty in this value.

$$\text{percentage uncertainty} = \frac{\text{uncertainty of instrument}}{\text{quantity measured with instrument}} \times 100$$

In this case, two measurements are made with the burette, one at the start and one at the end of the titration, so the equation is written:

$$\text{percentage uncertainty} = \frac{2 \times 0.05}{21.45} \times 100$$

$$\text{percentage uncertainty} = 0.47\%$$

> **! Common pitfall**
>
> Students often lose out on marks because they fail to appreciate that simply repeating a measurement will not improve the accuracy of that measurement — a more accurate piece of equipment is needed.

Worked example 4.5

Comparing the accuracy of different pieces of equipment

To determine the volume of carbon dioxide made when a given mass of a group 2 metal carbonate was reacted with dilute hydrochloric acid the apparatus was set up as shown in Figure 4.4.

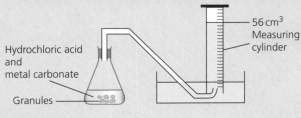

Hydrochloric acid and metal carbonate

Granules

56 cm³ Measuring cylinder

Figure 4.4

50 cm³ of the acid was placed in a conical flask. 0.45 g of powdered marble chips was weighed using a digital balance. A 100 cm³ measuring cylinder was filled with water and carefully inverted over the end of the delivery tube, to collect the gas produced in the reaction. The bung in the conical flask was removed the powder added quickly and the bung replaced to minimise any loss of gas.

The measuring cylinder in which the gas was collected was calibrated in 1 cm³ divisions and the balance was accurate to 0.01 g. Show by calculation which piece of apparatus contributed the biggest percentage error in this experiment.

Step 1: Write out the equation:

$$\text{percentage error} = \frac{\text{maximum error}}{\text{actual value measured}} \times 100$$

Taking the measuring cylinder first: the smallest gradation is 1 cm³, so the error range is ± half this value, or 0.5 cm³.

The volume measured was 56 cm³, so the percentage error is:

$$\frac{0.5}{56} \times 100 = 0.89\%$$

which should be rounded to 0.9%.

Step 2: Repeat for the balance — the maximum error is 0.01 g and the mass measured is 0.45 g, so the percentage error is (0.01/0.45) × 100 which is 2.22%. When comparing values, the value of the final answer can be no more accurate than the least accurate measurement, so 2.22% is given to one decimal place: 2.2%.

Step 3: Compare the answers. Although the balance is more accurate, since the mass weighed is so small, the percentage error will be greater.

Modifying procedures

Sometimes it is not the equipment that limits the accuracy of an experimental procedure, but limitations in the procedure itself. In this case, you need to consider whether modifying the procedure, using different equipment, would help. For example, in a calorimetry experiment where a spirit burner is being used to heat a given volume of water in order to determine the enthalpy of combustion, as shown in Figure 4.5, the main source of error is heat loss to the surroundings, which could be reduced by placing a heat shield around the apparatus.

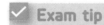
✓ Exam tip

Uncertainty is calculated in the same way as percentage error — by taking the value for the uncertainty for the piece of equipment and dividing by the measurement made with that piece of equipment, then multiplying by 100.

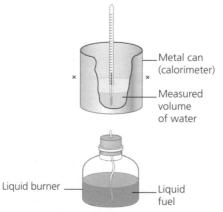

Metal can (calorimeter)

Measured volume of water

Liquid burner

Liquid fuel

Figure 4.5 Calorimetry set-up

One of the ways in which evaluative skills are tested is by asking you to consider the outcomes of specific changes. You have to make judgements on what the impact of each change on the result would be. For example, how what would happen if the reagent was changed to one with a higher or lower mass, or a volume was measured incorrectly As an A-grade student, you will recognise how the measurement is used and thus be able to predict the outcome of adjusting it in some way.

Application to the exam

In the exam you will be assessed on your ability to answer questions that cover the skills described in this chapter. The following worked example covers the higher-order skills of analysis (of why the cell does not work) and evaluation (of the relationship between concentration of silver ions and emf — electromotive force or cell potential of the cell).

Worked example 4.6

Judging the impact of changes

A student prepares a sample of copper(II) sulfate.$5H_2O$ by reacting a given mass of the insoluble copper(II) carbonate with sulfuric acid, filtering the solution to remove the excess solid and then heating gently before allowing to cool.

(a) The percentage yield of crystals is less than expected and the student suggests the following could be reasons for this. State and explain whether each statement is correct.

   (i) Not all of the copper carbonate reacted.

   (ii) Some of the water of crystallisation may have been lost when heating the solution.

   (iii) The crystals are still wet.

(b) If the student replaces the sulfuric acid with the same concentration and volume of hydrochloric acid, what would the impact be on the number of moles of crystals made?

Step 1: For part (a) take each statement in turn and decide whether it will have reduced the percentage yield of crystals:

   (i) The copper carbonate was in excess; it was the acid that was the limiting reactant, so this will not have reduaced the yield.

   (ii) The loss of water would reduce the mass of the crystals used to calculate the percentage yield, so this will have reduced the yield.

   (iii) In this case the mass of crystals would be higher than it should be, due to the water present. So the percentage yield would be higher, rather than lower.

Step 2: For part (b) writing out the equations for the reactions helps. As shown by the ratios in the following equations you need twice the number of moles of hydrochloric as you do sulfuric acid:

$$CuCO_3 + H_2SO_4 \rightarrow CuSO_4 + H_2O + CO_2$$

$$CuCO_3 + 2HCl \rightarrow CuCl_2 + H_2O + CO_2$$

The student will therefore make less product, because each mole of hydrochloric acid only produces one mole of hydrogen ions and so we would need twice the volume to produce the same number of moles of hydrogen ions as the sulfuric acid. The hydrogen ions are reacting to form water, so each mole of water formed requires two moles of hydrogen ions.

Worked example 4.7

A student sets up the electrochemical cell shown in Figure 4.6 but is unable to measure the electrode potential of the cell.

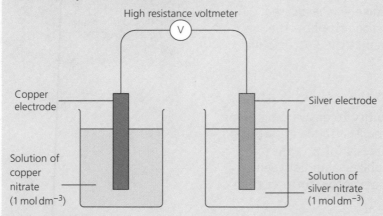

Figure 4.6

(a) Explain what is missing and why the cell does not work as it is currently set up. (3)

(b) Once the cell is working the results are recorded in Table 4.10. Describe what these results show. (2)

Table 4.10

$[Ag^+]/mol\,dm^{-3}$	Emf/V
1.0×10^{-4}	+0.25
3.3×10^{-4}	+0.28
1.0×10^{-3}	+0.31
3.3×10^{-4}	+0.34
1.0×10^{-2}	+0.38
1.0×10^{-1}	+0.43

(c) Explain why the change in the emf shown in the table above occurs and what impact increasing the concentration of the silver solution will have on the overall cell voltage. (5)

Table 4.11 The standard electrode potentials of some common metals

Metal ion/metal electrode	Standard electrode potential, E/V
$Li^+(aq) \mid Li(s)$	−3.03
$K^+(aq) \mid K(s)$	−2.92
$Na^+(aq) \mid Na(s)$	−2.71
$Al^{3+}(aq) \mid Al(s)$	−1.66
$Zn^{2+}(aq) \mid Zn(s)$	−0.76
$Fe^{2+}(aq) \mid Fe(s)$	−0.44
$Pb^{2+}(aq) \mid Pb(s)$	−0.13
$Cu^{2+}(aq) \mid Cu(s)$	+0.34
$Ag^+(aq) \mid Ag(s)$	+0.80

How to approach part (a)

The question is assessing your knowledge of how to set up a working electrochemical cell, so the first thing to do is to examine the diagram of the cell, which you are told does not currently work, and then identify each component. Start with the beaker on the left; does it contain an electrode in a solution of the metal ions? Is this also the case on the right? Will the circuit allow for the movement of electrons? Will it allow for the movement of ions?

The question asks you to explain what is missing and why the cell does not work. Once you have analysed the diagram, you will have identified that it is the salt bridge that is missing. In order to answer the question you need to know what the salt bridge is and does.

The salt bridge is a piece of porous material that is soaked in an ionic solution, such as potassium nitrate. Its purpose is to allow the passage of ions between the two solutions, while keeping them separate. This enables the charge in each beaker to remain balanced.

Sample answer

(a) The electrochemical cell is missing a salt bridge — a piece of filter paper soaked in potassium nitrate solution — which allows the movement of ions between the solutions to maintain the balance of charge. ✓

This is a simple, clear statement of what is missing from the diagram, what the salt bridge does and why it is important — it enables the movement of ions to maintain the balance of charge.

If this is not possible, then the opposite charges will build up in each beaker ✓ —positive in the beaker containing the copper electrode, where copper ions are formed, and negative in the beaker containing the silver electrode, where silver ions are removed ✓.

The answer explains that without a means for ions to move, the positive ions formed in the beaker containing copper will accumulate and the negative ions in the beaker containing silver will accumulate. This shows that the student understands that because copper is the more reactive metal it is forming ions, and that because silver is less reactive it is forming the metal, which leaves the nitrate ions in the solution in excess.

How to approach part (b)

From the fact that the concentration of silver is written in the first column of Table 4.12, you can deduce that this has been changed and the voltage for each concentration measured. Notice that this part is worth 2 marks, so a simple statement alone is unlikely to get both marks. In this situation you need to look for a numerical pattern using the data.

Sample answer

(b) As the concentration of the silver ions in solution increases, the emf becomes more positive. ✓

A statement is made linking the increase in concentration of silver ions with an increasingly positive emf. (Better to state it this way, rather than just say emf increases, because it could become more negative.)

A ten-fold increase in concentration results in an increase of 0.06 V. As you increase the concentration from 3.3×10^{-4} mol dm^{-3} to 3.3×10^{-3} mol dm^{-3}, the emf changes from 0.28 V to 0.34 V. ✓

The data in the table have been used to quantify the increase, and the example chosen been quoted.

How to approach part (c)

This question requires you to identify the reaction occurring in the silver half-cell. As you will see from Table 4.13, silver has a more positive standard electrode potential value and is less reactive (and less likely to form ions) than copper. This means that in the silver half-cell, the silver ions will be gaining electrons in a reduction reaction. This is represented as

$$Ag^+(aq) + e \rightarrow Ag(s)$$

Sample answer

(c) Silver has a more positive standard electrode potential (0.80 V) than copper (0.34 V) ✓ so silver is reduced and copper is oxidised ✓.

The electrochemical series is used to recognise that silver has a more positive value for the reaction involving the loss of electrons. This shows an understanding of the use of the electrochemical series as a measure of how likely a species is to lose electrons. The statement that silver is more likely to gain electrons and be reduced is awarded the mark.

The electrode potential for the cell is calculated as follows:

$$E_{cell} = E_{RHS} - E_{LHS} \checkmark$$

The correct equation is identified and the mark awarded.

where E_{RHS} is the value for silver, as this is the more positive and silver is being reduced. Under standard conditions this would be:

$$E_{cell} = 0.80 - (-0.34) \checkmark$$
$$= 1.14 \, V$$

The sign of the electrode potential for copper is reversed because copper is oxidised, gaining the mark.

If the E_{RHS} value becomes more positive, as the concentration of silver ions increases, the overall cell potential will become more positive; for the concentrations given, the value is less than the standard value. $\checkmark$

The answer recognises that increasing the value of E_{RHS} — the value from which you are subtracting the value of E_{LHS} — is going to make the cell potential more positive.

Take it further

The of University of York's chemistry department website has an interesting project page that includes information on different practical activities, such as clock reactions and the purity of aspirin.

You should know

> The key practicals that will be assessed by your particular exam board, including how to draw any diagrams associated with those practicals.

> How to process data, including how to select the information to use and the appropriate number of significant figures to quote.

> The difference between the limitations of the equipment (and how to quantify and compare errors) and the experimental technique.

5 Revision skills

Study skills

It is vitally important that the hard work of the previous years is not wasted through a failure to prepare adequately for the final assessment.

Revision is usually associated with preparing for exams; these will be concentrated at the end of the 2-year course. However, revision should not be left until this point. During your course, you will have tests (often as you complete a topic) and interim exams, so that you have some means of measuring your progress.

Learn as you go

As you progress through the course, there will be things you need to commit to memory. These include definitions and equations. How you do this is a matter of choice, but for the majority of people it takes time and repetition.

Activity

Take a topic from the specification — for example, atomic structure — and go through the specification identifying the definitions you need to learn. Produce a key card for each definition (blank white postcards are good for this) — write the term you are defining on one side and the definition on the other. Keep the pack of definition cards for each topic separate (you could key code them with the specification reference) and learn them.

For example:

OCR 2.1.1.c	
Relative atomic mass	The weighted mean mass of an atom of an element compared with 1/12 of the mass of an atom of carbon-12

Summarising information

In addition to memorising key facts, it is also useful to produce your own summaries of information. These can be simple flow diagrams — for example, to describe the key points about mass spectrometry — or they can be more complex summaries of different types of mechanism or synthetic pathways.

Synthetic pathways

Where you have to learn reaction sequences, create posters as you go. On a large sheet of paper (e.g. A3) list the functional groups from the specification. Can you convert one to another? If so, add in reagents and conditions (use a different colour for each). Add the formulae for the products formed.

Start this when you first begin organic chemistry and add to it as you progress through the course. Use these posters for reference when answering questions or devising reaction sequences. They can also be used as a revision aid, where you cover sections and try to recall what is obscured. Practise writing out the mechanisms, and look for similarities and differences.

> **Activity**
>
> Produce a simple flow diagram to show the stages in time-of-flight mass spectrometry.

> **Activity**
>
> Develop the example in Figure 5.1 by adding other organic chemicals or different members of the homologous series. 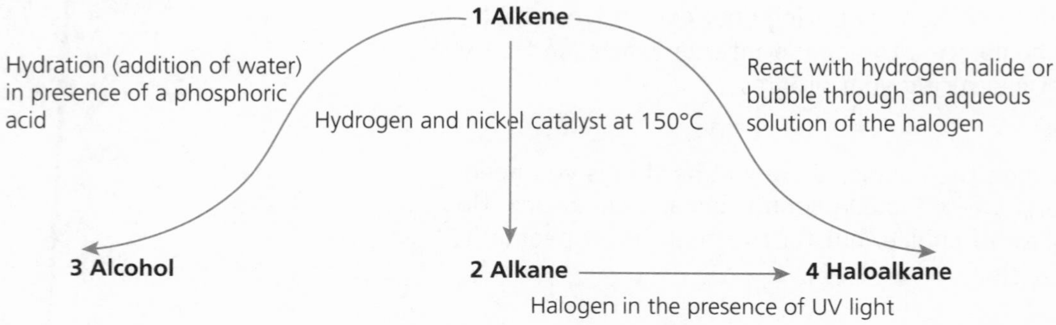
>
>
> Figure 5.1

Another way of learning reactions is to make a set of cards. Put the starting material on one card, what it will react with on another (you could also include reaction conditions or these could go on a different card) and the product on another card. If you use a table template the cards will be of a similar size, so can be cut out and matched up (plus you will have the template for reference to check your answers against). Table 5.1 shows an example.

Table 5.1

Starting compound	Reagents	Conditions	Product	
Benzene	$HNO_3(l)$	50°C + conc. sulfuric acid catalyst	Nitrobenzene	
Acid anhydride E.g. ethanoic anhydride $(CH_3CO)_2O$	Alcohol Ethanol CH_3CH_2OH		Ester + carboxylic acid Ethyl ethanoate $(CH_3COOCH_2CH_3)$ + ethanoic acid (CH_3COOH)	

Application to the exam

Practice questions

Ultimately you will be tested on what you know through a series of questions. The more familiar you are with how your exam board asks the questions, the better you will do.

Multiple-choice questions

Multiple-choice questions feature in many of the specifications on at least one of the papers. They can easily take up more time than they should and so getting plenty of practice at answering these types of question and refining your technique is important. Multiple-choice questions are designed so that the most common wrong answers are included in the selection from which you can choose; these can include misconceptions and the most easily made mistakes. You might like to try to predict what these would be too.

When answering multiple-choice questions, try to work out the answer before you look at the responses; then look to see if it is there. That way you are less likely to pick a common error response. There are different types of multiple-choice question, as the following worked examples show.

Worked example 5.1

Where there is only one correct answer

If 100 g of each of the following compounds is dissolved in water, which of the solutions will have the highest pH value?

A $NaOH$

B $Ca(OH)_2$

C $Mg(OH)_2$

D $CaCO_3$

In this question, there are a number of distractors. The first thing to consider is that the group 2 hydroxides will give twice as many hydroxide ions per mole of compound than the group 1 hydroxides.

It is tempting to think that because $Mg(OH)_2$ has a lower relative formula mass than $Ca(OH)_2$, then this will give a greater number of hydroxide ions (for the same mass), so the answer would be C; however, $Mg(OH)_2$ is less soluble than $Ca(OH)_2$, so the correct answer is B.

Worked example 5.2

Where the answer is based on which responses may be correct

As a result of the delocalisation of electrons:

(i) benzene undergoes addition rather than substitution reactions

(ii) propanoic acid is a stronger acid than propanol

(iii) phenylamine is a stronger base than ethylamine

(iv) ethylamine is a stronger base than ammonia

Select the correct answer from the responses below:

A   (i), (ii) and (iii)

B   (i) and (iii)

C   (ii) and (iv)

D   (iv) only

The best way of tackling these questions is to identify which statements are true, and then to look at the combinations offered.

(i) False — benzene undergoes substitution reactions not addition ones.

(ii) True — propanol forms a neutral solution in water.

(iii) False — ethylamine is a stronger base than phenylamine.

(iv) True — ethylamine is a stronger base than ammonia due to the inductive effect of the atoms attached to the nitrogen, which accepts the proton.

Having identified that (ii) and (iv) are correct, you can now see that the correct answer is C.

An A-grade student will anticipate and avoid, or at least recognise, the common wrong answers.

> **! Common pitfall**
>
> It is easy to spend too long on multiple-choice questions. A very rough rule of thumb is a mark per minute, and it is a good idea to leave the multiple-choice questions to the end. If there are 20 questions, allow 25 minutes at the end of the exam. Work through those you are able to do first and leave any you find tricky to the end. Come back to these after you have completed the rest and have another go. If all else fails, make an educated guess.

Planning answers

There are different ways to practise answering questions:

→ You can answer them under timed conditions, without reference to your notes.

→ You can do as much as you can without reference to your notes, then look up any areas you are unsure of.

→ You can plan an answer by making bullet point lists of the key parts and then, checking these, write the answer in the time allowed.

→ You can pick out the parts you need to practise and just tackle those.

→ Or, most likely, you can do a combination of the above.

It is useful in exams to list the key points of longer, unstructured questions because this enables you to write a logical and sequential response.

One possible way of planning your question practice is to tabulate all the different papers and then tick them off as you complete them; perhaps also recording your score as you go. An alternative method would be to list the different topics and then to record the scores you achieve in any assessed work you do on each topic. You should convert any marks to a percentage for easier comparison. Either way, you are tracking where you need to improve.

Getting organised

Whether this is for the A-levels themselves or for interim tests and exams, it is important to know what you will be tested on, what will be on each paper, the format of the paper and the length of the exam.

As an A-grade student, you will understand the need to act on the feedback you get throughout the course and to be aware of your strengths and areas for improvement. You need to use the feedback from every piece of work as a diagnostic tool to help you direct your efforts most effectively.

To plan your revision, use a calendar that gives you a clear view of how much time you have between now and the exams. Add in the dates of your chemistry exams and any other exams you will be sitting. Now block off any times when you have commitments that mean you are not able to revise (it is important to have some time for relaxation too). Table 5.2 gives an example.

Table 5.2 Sample revision timetable

Monday	Tuesday	Wednesday	Thursday	Friday	Saturday	Sunday
		1 Biol am Chem pm	2 Biol am Maths pm	3 **Biol paper 1 am** Chem pm	4 Maths am Chem pm	5 Chem am Maths pm
6 **Maths paper 1 am** Biol pm	7 Chem am Maths pm	8 **Chem Paper 1 am** *Play squash* Biol pm	9 Maths am Chem pm	10 **Maths paper 2 am** Biol pm	11 Chem am *Cinema pm*	12 Chem am Maths pm
13 Biol am Chem pm	14 **Chem paper 2 am** Maths pm	15 Biol am Maths pm	16 Biol am Maths pm	17 **Biol paper 2 am** **Maths paper 3 pm**	18 *Shopping* Chem pm	19 *Fathers' day — lunch out* Biol pm
20 Biol am Chem pm	21 **Biol paper 3 am** Chem pm	22 Chem am *Play squash*	23 Chem am Chem pm	24 **Chem paper 3 am** *Last exam!* *Celebrate!!*	25	26
27	28	29	30			

Once you have allocated which subjects to cover on each day, you then need to break this further into topics. For example, for the week commencing 20 June, supposing paper 3 is the one in which your exam board assesses practical skills. For this you could produce a weekly plan like the one in Table 5.3.

Table 5.3 A weekly revision plan

Monday 20 pm	Revisit essential practical techniques
Tuesday 21 pm	Go over calculations associated with practicals, including error determination, titration, calorimetry
Wednesday 22	Revise the reagents and results of qualitative tests
Thursday 23	Revisit any of the above — as necessary

Knowing when your exams are will give you an idea of when you need to cover each topic by.

Some students find it helpful to save a set of papers to do in their entirety under exam conditions, like another mock. Others prefer to focus on a particular topic and do as many questions on that topic as they can. Ideally, you will do both. What is vital, though, is that you do as many questions as you can and that each time you review your performance.

Last-minute preparation

Do not be tempted into last-minute revision; this is likely to do more harm than good, as it will make you feel anxious and you will be less likely to do yourself justice. Check that you have what you need in terms of equipment, and remember to use the data sheet.

In the exam

Before you get into the exam, spend a few minutes allocating your time. If you have multiple-choice questions, work out how much time you need to leave for these — when in the examination hall jot down the timings on the front of your paper. You do not have to start at the beginning of the paper, or do the questions in order, but do check both sides of each page and make sure that you answer *all* of the questions. If you are leaving one out to go back to, fold over the page as a reminder.

Read all the way through a question before you answer it. This has two benefits: it often shows how a topic develops through the question, which can be helpful when you are planning your answer; and it enables your subconscious to start recalling the information you need.

When you have completed the paper, check your calculations and that you have answered all parts of all of the questions, but do not be tempted to make a lot of changes; your first answer is often correct.

Take it further

A useful website with lots of suggestions on different ways to structure your work, as well as information on the importance of practice, and guidance on the step up to A-level, is: www.alevelmindset.com.

You should know

> **That to get achieve the top grade, you need to start preparing from the first day of the course and to establish the habits that have been highlighted in this chapter.**

> **The importance of adopting a diagnostic attitude, so that you track the grades you achieve in each piece of work and you know your strengths and weaknesses in terms of knowledge and skills.**

> **How to check each piece of work before handing it in, correcting any errors you spot, so that this becomes second nature.**

> **To regularly review your work, making summaries of the information where appropriate.**

> **How to look for links and similarities, as well as differences, across topics.**

Exam board focus

The course you are following and the awarding body are likely to have been chosen for you by your school.

Activity

Use the exam board website for the qualification you are taking to find out the key information mentioned above. (You will probably need to refer to the specification.)

Copy and complete Table 6.1.

Table 6.1

Exam board and specification reference number	
Number of papers	
For each paper: What content is tested? How long is the paper? What type of questions are there? How many marks is the paper worth? What percentage of the overall A-level does it account for?	
How many assessment objectives are there and what are they?	
How is practical work assessed?	

What you need for an A

The difference between...

Many students lose marks because they do not know or remember the work covered in the first year of their course, as well as the more recent A2 work.

A B-grade student	An A-grade student
• Has notes that are clearly labelled, but where topics may be interspersed • Highlights key definitions in their notes and learns these for exams • Revisits topics in preparation for tests and exams • Knows which topics they find the hardest	• Has notes organised into sections that reflect the topic areas of the specification • Has definition cards that complement the topics of the specification and knows these definitions • Regularly revisits topics that have been covered previously • Focuses more effort on the topics they find hardest

Grade boundaries are set by the awarding bodies, using statistical methods and a comparison with previous cohorts, as well as the predictions of achievements for the current cohort.

To prepare fully for the requirements of your own exam board, use past papers alongside the mark schemes and the examiners' reports. Exam boards produce a lot of materials to support the teaching of their specification, and apart from the most recent exams and assessment materials, most of these are readily available on the website.

Activity

Go to the website for your exam board and look at what material is available. How many past papers or assessment materials are there? What other resource do you think you might find helpful? How could you use the materials here?

The different specifications

Spotlight on AQA

AQA offers AS and A2 as separate, stand-alone qualifications. Students choosing to do A2 sit three separate 2-hour papers at the end of Year 13. There is a separate practical endorsement and practical questions are incorporated into all three papers.

Spotlight on CCEA

CCEA still includes a practical examination and it also still credits the AS component towards the A2 qualification. So students must take six written exams for the full A-level qualification, whereas with the other boards, now that the AS and A-level qualifications are decoupled, students can take a minimum of three written papers.

Spotlight on Edexcel

Edexcel offers AS and A2 as separate, stand-alone qualifications. Students choosing to do A2 sit three separate papers, two of which are 1 hour 45 minutes, with the third being 2 hours 30 minutes. The practical endorsement is separate from the examinations.

Spotlight on OCR

OCR offers two different specifications — OCR A and OCR B. The OCR B specification is concept led rather than content led. What this means is that chemistry is taught through a series of different storylines or scenarios, where the chemistry is placed in context.

This places a greater emphasis on problem solving and investigative practical work, and ideas are introduced in a spiral manner, so that a topic is introduced in Y12 and then revisited in Y13.

Spotlight on WJEC

The WJEC qualification, which is not available in England, includes a practical examination and it still credits the AS component towards the A2 qualification. There are five units in total, one of which is the practical examination, which is taken in the final year of the course.

The Eduqas qualification from WJEC is available for use in England and mirrors the format of the other exam boards in terms of the practical endorsement and the decoupling of the AS and A2 exams.

Activity

Go to your exam board's website and find the SAMs (specimen assessment materials) available; download these and work through them to familiarise yourself with what the examiners expect for the higher grades. What general principles can you learn? Try to apply these to your own answers.

Activity

Download a copy of one of the papers. Now go through and look at the structure of the questions — not just whether they are broken down into shorter sections or longer answers, but also the skills they are assessing. See if you can identify the questions that just involve the recall of information, those that assess maths skills, those based on the practicals and the ones that require you to apply your knowledge. Highlight the different sorts of questions using different colours.

Then look carefully again at where the questions give you guidance, this might be in the number of significant figures required in an answer, or it might be 'with reference to a table', or similar. Next to each question add the topics that are covered.

You should now have a much better idea of how papers are compiled, in terms of content and coverage of the assessment objectives.

You should know

> **The topics covered on each paper, the style of the questions and how the exam board awards marks.**
> **What is on the website for your board, by checking this regularly to take advantage of any new material that has been added.**
> **That in order to achieve the highest marks and hence the top grade, you should be very familiar with your exam board's assessment structure.**

Quantitative skills answers

Here you will find answers for the activities in chapter 1 (Quantitative skills). Answers are not included for all of the activities suggested. For example, those that involve researching information could vary depending on which exam board you are following, and so are not exemplified.

Pages 10–11, Table 1.1

Units

(a) $H_2(g) + I_2(g) \rightleftharpoons 2HI(g)$ No units
(b) $N_2(g) + 3H_2(g) \rightleftharpoons 2NH_3(g)$ Units $mol^{-2} dm^6$

Decimal places

$$pH = -\log_{10}[H^+]$$

$$= -\log_{10} 0.136$$

$$= 0.86646$$

$$= 0.87$$

Standard form

(a) 4.5×10^{-4}
(b) 3.75×10^5

Ratios, fractions and percentages

(a) Moles of 2-hydroxybenzoic acid $= \dfrac{2.00}{(7 \times 12.0) + (6 \times 1.0) + (3 \times 16.0)}$

Multiply the result by $(9 \times 12.0) + (8 \times 1.0) + (4 \times 16.0)$.

Answer: 2.61 g

(b) 66.7% of 2.61 = 1.74 g

(c) If 51.6 g is oxygen and 9.1 g is hydrogen, then 100 − (51.6 + 9.1) = 39.3 g is carbon.

Dividing the mass of each element by its relative atomic mass gives:

51.6/16.0   9.1/1.0   39.3/12.0
3.225   9.1   3.275

Dividing the numbers by the smallest value gives:

1   2.82   1.02

Rounding these gives:

1   3   1

So the empirical formula is CH_3O.

Estimating

If the concentration of the alkali is the same, you would expect titres of around $50\,cm^3$, because each mole of acid requires 2 moles of alkali for neutralisation. If the concentrations are the same then the volume of the alkali will be double that of the acid.

Means

Titration	1	2	3	4	Mean
Titre/cm³	22.00	22.35*	22.20	22.45*	**22.40**

*Concordant titres are used to find the mean.

Uncertainty

$$\frac{1}{23} \times 100 = 4.35\%$$

Equations

$$K_a = \frac{[H^+]^2}{[HA]}$$

So:

$$[H^+] = \sqrt{K_a[HA]}$$

$$= \sqrt{1.6 \times 10^{-4} \times 0.01}$$

$$= 1.265 \times 10^{-3}$$

$$pH = -\log_{10} 1.265 \times 10^{-3}$$

$$pH = 2.90$$

Logarithms

(a) −0.602

(b) −1.39

(c) $x = 10^{0.32} = 2.09$

Graphs

The graph represents a second-order reaction.

Geometry and trigonometry

Page 14, Activity

(a) 376 000 Pa

(b) 42.0 g

(c) 0.00732 mol

(d) 5760 J

Page 15, Table 1.2

Symbol	Meaning
⇌	A reaction that is reversible because it can proceed in both directions
≥	Greater than or equal to
≈	Approximately equal to
<	Less than
~	Approximately
∝	Proportional to

Page 15, Activity

(a) To get V on its own on the left-hand side of the equation, we divide both sides by p:

$$V = \frac{nRT}{p}$$

(b) To make c the subject of the equation, multiply both sides by c^2, giving:

$$K_c c^2 = ab$$

Divide each side by K_c, giving:

$$c^2 = \frac{ab}{K_c}$$

Take the square root of each side, giving:

$$c = \frac{\sqrt{ab}}{K_c}$$

(c) To get M_r, multiply both sides by M_r, giving:

$$n \times M_r = m$$

Then divide both sides by n, giving:

$$M_r = \frac{m}{n}$$

(d) To get T, first add $T\Delta S$ to each side, giving:

$$\Delta G + T\Delta S = \Delta H$$

Then subtract ΔG from each side, so:

$$T\Delta S = \Delta H - \Delta G$$

Then divide each side by ΔS, giving:

$$T = \frac{\Delta H - \Delta G}{\Delta S}$$

Page 15, Activity

Rate constant, k:

rate $= k[A][B]$

where the rate is first order with respect to both A and B, and where the only thing being changed is the concentration of the reactants.

Equilibrium constant:

For a given reaction $aA + bB \rightleftharpoons cC + dD$:

$$K_c = \frac{[C]^c[D]^d}{[A]^a[B]^b}$$

Equilibrium constant for a homogeneous reaction involving gases:

For the reaction $aA(g) + bB(g) \rightleftharpoons cC(g) + dD(g)$:

$$K_p = \frac{p(C)^c p(D)^d}{p(A)^a p(B)^b}$$

where p is equal to the partial pressure of the gas, calculated from the mole fraction and the total pressure.

Acid dissociation constant, K_a:

For a weak acid, HA (which is not fully dissociated in solution):

$$K_a = \frac{[H^+][A^-]}{[HA]}$$

This is sometimes adapted for indicators, in which case the 'a' is replaced by 'ind':

$$K_{ind} = \frac{[H^+][Ind^-]}{[HInd]}$$

Page 18, Activity

(a) In this graph the variable plotted on the y-axis is proportional to that on the x-axis. This can be written as $y \propto x$, which can also be written as $y = mx + c$ where m is the gradient of the line and c is the intercept on the y-axis. This graph could represent a reaction that is first order with respect to a particular reagent.

(b) This graph shows that the variable plotted on the y-axis is proportional to the variable on the x-axis raised to a power of more than 1. This can be expressed as $y \propto x^n$, where $n > 1$. The graph could represent a reaction that is second order with respect to a reagent. (Curves are typically obtained when showing the change in solubility with temperature.)

(c) The variable plotted on the y-axis is independent of that plotted on the x-axis. This can be written as $y \propto x^0$, which means that y is constant and that for any given value of x, y stays the same. This type of graph is seen if a rate is zero order with respect to a given reagent, because changing the concentration of that reagent will not change the rate of the reaction.

Page 20, Activity

Page 21, Table 1.3

No. of bond pairs	No. of lone pairs	Shape	Bond angle/°	Example
2	0	Linear	180	$BeCl_2$
3	0	Trigonal planar	120	BF_3
3	1	Pyramidal	107	NH_3
4	0	Tetrahedral	109.5	CH_4
2	2	Angular (bent)	104.5	H_2O
5	0	Trigonal bipyramidal	90 and 120	PF_5
6	0	Octahedral	90	SF_6
4	2	Square planar	90	XeF_4

Other examples may be given.

Notes

Aiming for an A in A-level Chemistry